T0091499

Physics on Ultracold Quantum Gases

Peking University–World Scientific Advanced Physics Series

ISSN: 2382-5960

Series Editors: Enge Wang *(Peking University, China)*
Jian-Bai Xia *(Chinese Academy of Sciences, China)*

Peking University-World Scientific Advanced Physics Series

Vol
8

Physics on Ultracold Quantum Gases

Editors

Yongjian Han

Wei Yi

Wei Zhang

University of Science and Technology of China, China

北京大学出版社
PEKING UNIVERSITY PRESS

World Scientific

Published by

World Scientific Publishing Co. Pte. Ltd.
5 Toh Tuck Link, Singapore 596224
USA office: 27 Warren Street, Suite 401-402, Hackensack, NJ 07601
UK office: 57 Shelton Street, Covent Garden, London WC2H 9HE

Library of Congress Cataloging-in-Publication Data
Names: Yi, Wei, 1977–   author. | Han, Yongjian (Physicist), author. |
    Zhang, Wei (Physicist), author.
Title: Physics on ultracold quantum gases / Wei Yi, Yongjian Han, Wei Zhang
    (University of Science and Technology of China, China).
Description: New Jersey : World Scientific, 2018. | Series: Peking University-World Scientific
    advanced physics series ; v. 8 | Includes bibliographical references.
Identifiers: LCCN 2018014012| ISBN 9789813270756 (hardcover : alk. paper) |
    ISBN 9813270756 (hardcover : alk. paper)
Subjects: LCSH: Cold gases. | Low temperatures. | Superfluidity. | Quantum statistics.
Classification: LCC QC175.16.C6 Y5 2018 | DDC 536/.56--dc23
LC record available at https://lccn.loc.gov/2018014012

British Library Cataloguing-in-Publication Data
A catalogue record for this book is available from the British Library.

Copyright © 2019 by Wei Yi, Yongjian Han and Wei Zhang

The work is originally published by Peking University Press in 2014.
This edition is published by World Scientific Publishing Company Pte Ltd by arrangement with
Peking University Press, Beijing, China.

All rights reserved. No reproduction and distribution without permission.

For any available supplementary material, please visit
https://www.worldscientific.com/worldscibooks/10.1142/10999#t=suppl

Typeset by Stallion Press
Email: enquiries@stallionpress.com

Preface

This book draws from the graduate courses *"Physics of Ultracold Atomic Gases"* (since 2010) and *"Cold Atom Physics"* (since 2011) from the Renmin University of China and the University of Science and Technology of China, respectively, which, for the past few years, have served as elementary introductions to cold atomic gases for interested students. While the main textbook for both courses is the popular *"Bose–Einstein Condensation in Dilute Gases"* by Pethick and Smith, we decided that it would be in our best interests, as well as our students, to have a textbook of our own, so that they may have the convenience of a complementary reference, modest as it is, when confused in class. For that purpose, we have intended from the start to be pedagogical and not compromising. Whether this is indeed the case remains to be seen.

All the authors contributed extensively to the writing of the book: Part III is written by Yongjian Han, the Introduction and Part II are written by Wei Yi, and Part I is written by Wei Zhang. As such, the authors are listed in alphabetic order. We are grateful to Jiansong Pan for proof-reading some of the chapters and to Zhen Wang for making some of the figures in Part III. We also acknowledge all the students who have taken the abovementioned courses, whose brilliant feedbacks proved to be immensely helpful. We thank Xiaohong Chen from the Peking University Publishing House, whose continuous support made this project possible. Finally, we apologize for the inevitable mistakes lurking out there and hope that we may have the chance to improve in the future.

Beijing,   *Yongjian Han*
July 2014   *Wei Yi*
                                                              Wei Zhang

Contents

1
Introduction

Almost twenty years have passed since the first experimental observations of Bose–Einstein condensation (BEC) in ultracold atomic gases [1–3]. In these seminal experiments, up to $10^4 \sim 10^5$ neutral bosonic atoms of alkali were trapped using lasers and magnetic fields and were cooled down to temperatures below the microkelvin range, at which point the collective behavior of the gas becomes quantum mechanical. The realization of BEC in dilute atomic gases turns a new page in our understanding of Nature and provides us with a versatile platform, on which previously unsolved physical problems can be studied and many novel ideas can be tested. Coupled with other recently developed techniques, such as the Feshbach resonance, the optical lattice potentials and the synthetic gauge fields, and ultracold atomic gases — both bosonic and fermionic — are playing an increasingly important role in various fields of research, including quantum simulation, quantum computation and precision measurement, to name a few.

For a systematic introduction of BEC, we refer the readers to the excellent book by Pethick and Smith [4], where the authors also cover topics like cooling and trapping of cold atoms that are critical for the realization of BEC. The purpose of this book is to serve as a modest introduction to the more recent progresses, such as the Fermi condensate, polarized Fermi gas, synthetic gauge fields and quantum simulation with optical lattice potentials, etc. We will also cover the basic theoretical framework under which physical processes such as two-body scattering, pairing superfluidity of fermions and so on, are modeled. Before doing so, let us first review, from a historical perspective, the realization of condensation of both bosonic and fermionic atoms in ultracold atomic gases.

The study of BEC dates back to 1924, when Bose introduced a new way of counting the microscopic states of the radiation field [5]. Using this method, Bose was able to re-derive Planck's formula for the black-body radiation in his seminal paper. Einstein later extended the approach to treat massive indistinguishable particles that obey the same statistics [6]. The resulting Bose–Einstein distribution leads to the striking conclusion that a majority of the particles would occupy a single quantum state with the lowest energy at low enough temperatures. Here, the occupation of the ground state is closely related to the system's ability to accommodate particles in the excited states, which increases with temperature. As a result, Bose–Einstein condensation should occur in a system with a low enough temperature and a high enough number density.

Physically, as the temperature is lowered, the thermal de Broglie wavelength of the particles in the system increases. Condensation occurs when the thermal de Broglie wavelength is on the order of the inter-particle separation. At this point, the wave packets of different particles overlap and interfere with each other to form a larger wave packet common to all particles in the condensate. A more detailed calculation shows that the condition for condensation in a free space of three dimensions is $\lambda n^{1/3} \sim 2.612$, which is consistent with the hand-waving argument outlined above. Here, n is the number density and λ is the thermal de Broglie wavelength. Considering the expression $\lambda = (2\pi\hbar^2/mk_BT)^{1/2}$, this argument is consistent with the previous analysis that condensate occurs with high density and low temperature. One often regards $\lambda n^{1/3}$ as the phase-space density [7]. To achieve condensate experimentally is to look for and implement a system whose phase-space density can go beyond 2.612.

Although the physics behind the theory is elegantly simple, it turned out to be quite difficult to realize a well-behaved BEC experimentally. To increase phase-space density often requires a high number density, which typically leads to the solidification of the system. The only exception is liquid ^{4}He, which remains a liquid at the lowest temperature. Regardless, whether it be in solids or in liquid helium, the interaction is typically quite strong and cannot be simply neglected. With interaction, Einstein's theory must be modified and it turns out that a strong interaction leads to the so-called depletion of the condensate, i.e., the process in which interaction-induced excitations make particles leave the condensate. In fact, although it has been suggested in 1935 that the superfluidity of liquid helium is related to BEC, the population of the ground state in a typical ^{4}He superfluid is only ~10% of the total particle number [8]. Furthermore, due to the strong interaction, the properties of the system cannot be characterized analytically. Hence it seems that the only way to achieve a weakly-interacting BEC is with a low number density. However, at first glance, there are many questions to be answered: whether a system with such a low density is stable, how does one trap the system in space and cool the system to the required temperature, which given the low density of the system, would be lower than any known temperatures in the whole Universe.

Soon, people realized that these problems can be solved in a dilute atomic gas. With a typical density of $n \approx 10^{12} \sim 10^{15}/\mathrm{cm}^3$, these systems are usually metastable. Yet, so long as the lifetime of the dilute gas is much longer than the time required for thermal equilibrium, it is possible to realize and observe Bose–Einstein condensates with weak interactions under typical experimental conditions. Due to the low particle density, the interatomic interaction is typically rather weak in these systems, such that they can be well characterized in a perturbative fashion. The catch, however, lies in the extremely low temperature required. In this case, for the phase-space density to reach unity, the temperature should be below the microkelvin range. There is also the question of how to hold the gas in space. Containers are obviously out of the question. Hence, the key to BEC in a dilute atomic gas seems to lie in the cooling and trapping of the gas. Indeed, in hindsight, the three key developments that eventually paved the way for the realization of BEC in dilute gases are: laser cooling, trapping with magnetic field or laser and evaporative cooling [9].

Early experimental efforts toward BEC focused on dilute gases of spin-polarized hydrogen atoms, which feature weak attractive interaction and are metastable against molecule formation in a gaseous form. It was proposed that BEC as well as superfluidity can be realized in such a system [10]. Although the BEC of spin-polarized hydrogen atoms was not realized until 1998, that is, after the BEC in alkali atoms, many important techniques had been first developed for hydrogen atoms, e.g., trapping with magnetic field and evaporative cooling. For a dilute hydrogen atomic gas, the atoms are initially cooled cryogenically, which requires the gas to be in contact with a cold surface. This limits the achievable density of the gas owing to the interaction of atoms with the container wall. To avoid contact with the container, it is preferable to hold the atomic gas in place by a magnetic field. These ideas were entertained at MIT by Greytak and Kelppner, and led to the invention of magnetic trapping.

A revolutionary advance took place in the 1980s with the advent of laser cooling. Though not applicable to hydrogen atoms, laser cooling has the potential to cool a whole spectrum of atoms, e.g., alkali atoms, rare-earth atoms, alkaline-earth atoms etc., from room temperature down to hundreds of microkelvins. This is followed by the invention of magnetic-optical traps, in which the atoms can be cooled further via the Sisyphus cooling mechanism and reach the range of tens of microkelvins. There is however still a final gap to overcome before the condensation can take place. In the following years, there were many brilliant proposals for the so-called sub-recoil cooling, which aims to go below the temperature limit of previous laser cooling schemes. The practical limit of laser cooling is set by the scattering of atoms by laser, a necessary process in the cooling scheme. Thus the task of cooling further with laser seems insurmountable. The solution lies in the introduction of a second cooling stage, the evaporative cooling. Originally devised to cool spin-polarized hydrogen atoms, the evaporative cooling selectively removes the hottest atoms in the trap and allows the rest to thermalize. As a result, the number density increases while the temperature decreases. Thus, evaporative cooling is able to overcome the last several orders of magnitude in the final climb of the phase-space density and is an essential step in realizing BEC in dilute atomic gases. In 1995, E. Cornell and C. Wieman at the University of Colorado Boulder, R. G. Hulet at Rice University, and W. Ketterle at the Massachusetts Institute of Technology (MIT) were the first individuals to successfully combine laser cooling with evaporative cooling and observe BEC in dilute gases of atoms. Decades of heroic experimental endeavor eventually paid off.

Soon after the realization of BEC in dilute gases of bosonic atoms, people started to think about bringing fermionic atoms into quantum degeneracy. While it is possible for bosonic atoms to occupy the same state, for fermionic atoms, a given quantum state can only accommodate at most one atom at most due to the Pauli blocking. As a result, the atoms in a Fermi gas at zero temperature occupy all available low-energy states, thus forming a Fermi sea with a sharp edge in momentum space. However, in the presence of attractive interatomic interactions, the Fermi sea becomes unstable and fermions at the surface of the Fermi sea pair up to form the so-called Cooper pairs. The Cooper pairs are composite bosons and may condense to form a BEC at low temperatures [11]. This is the physical picture behind the famous Bardeen–Cooper–Schieffer (BCS) theory proposed in the 1950s to explain superconductivity [12] in metals. It is therefore interesting if one can also prepare such a Fermi condensate of composite bosons in an ultracold atomic Fermi gas.

The problem again lies in cooling. According to the BCS theory, the critical temperature at which the Fermi condensate emerges in an attractively interacting Fermi gas decreases exponentially with the interaction strength. In a dilute gas, the interaction is typically very weak, hence it seems impossible to reach the condensation temperature in the first place. In 1993, E. Tiesinga, B. Verhaar and H. Stoof proposed that the interatomic interactions can be tuned via the Feshbach resonance technique [13, 14], which was first discovered by H. Feshbach in 1958, in the context of nuclear physics [15]. In a Feshbach resonance, the two-body scattering length can be tuned by adjusting the external parameters. Microscopically, the effective s-wave scattering length can be smoothly adjusted from zero to infinity and from negative to positive by tuning the relative energy difference between a two-body quasi-bound state and the threshold of the two-body scattering continuum. In an ultracold atomic gas, such an energy difference can be tuned via either the external magnetic field or the laser field. The Feshbach resonance was observed in ultracold atomic gases in 1998 [16–18], and has since proven to be one of the most powerful tools in ultracold atoms, as it can lead to strong correlations in a dilute atomic gas.

For an ultracold Fermi gas, this implies a dramatically increased critical temperature and the possibility of observing Fermi condensate in ultracold Fermi gases under practical experimental conditions. In 2003, condensation of fermion pairing states was observed in ultracold gases of ^{6}Li and ^{40}K [19–21]. Furthermore, by tuning the interaction strength, it was possible to demonstrate the crossover between a tightly bound molecular BEC of composite bosons and a condensate of fermion pairs with long-range correlation. This so-called BCS-BEC crossover had previously been studied theoretically in the context of condensed matter physics as a possible scenario for high T_c superconductivity [22]. With the advent of ultracold Fermi gases, it is now possible to experimentally investigate the rich physics of the BCS-BEC crossover in a completely different physical system.

Another critical tool to achieve strong correlation in a dilute atomic gas is the implementation of an optical lattice [23]. With atoms arranged into a spatially periodic lattice potential, it is possible to experimentally simulate the lattice models that are important in solid-state physics. Together with tools like the Feshbach resonance, the idea of quantum simulation has significantly changed our perception of a dilute atomic gas. With tunable interactions and a clean environment, and with both bosons and fermions at our disposal, the ultracold atomic gases have become an ideal platform to study strongly interacting strongly correlated systems.

The remainder of this book contains three parts. In Part I, we provide a brief introduction of the necessary background knowledge including atomic physics, the quantum description of light-matter interaction, trapping and cooling of ultracold atoms, and the physics of two-body interaction, which have crucial impacts on the many-body properties of a dilute atomic gas. We then discuss in detail the properties of a dilute Fermi gas in Part II. For the final sections in Part III, we discuss the recent quantum simulation schemes for cold atoms loaded into an optical lattice. In the following, we will give a more detailed introduction to the contents of the various parts.

Part I consists of five chapters. In Chapter 2, we give a very brief review of atomic physics, including the level structure and the Zeeman splitting. We then cover the quantum description of light-atom interaction, i.e., a minimal version of quantum optics in Chapter 3. Then in Chapter 4, we discuss typical trapping and cooling schemes for ultracold atoms. In Chapter 5, we review the interaction potential between two atoms, before introducing the two-body scattering physics as well as the concept and application of effective potentials. We devote Chapter 6 to the description of the Feshbach resonance, which is of central importance to many of the topics that we will discuss later in the book. The topics covered in this part serve as background knowledge for the rest of the book. Interested readers may look up more detailed reviews in the reference sections at the end of each chapter.

Part II consists of five chapters. In Chapter 7, we review the BCS theory for weakly interacting Fermi gases. We then introduce the BCS-BEC crossover theory as a natural extension of the BCS description in Chapter 8. We discuss in detail the crossover problem on the mean-field level, on which many important physical understandings of the crossover process can be established. In Chapter 9, we introduce several beyond-mean-field treatments of the BCS-BEC crossover problem. These many-body calculations provide quantitative corrections to the mean-field results, which may then be compared with experimental measurements. In Chapters 10 and 11, we review the interesting topics of polarized and spin-orbit coupled atomic gases, respectively. Both of these topics have been of great theoretical and experimental interest in recent years, and both can be understood qualitatively well on the simple mean-field level.

Part III focuses on quantum simulation with cold atoms in an optical lattice potential. We start by briefly reviewing the implementation of optical lattices in cold atomic gases, as well as the band structures in an optical lattice in Chapter 12. With the basic understanding of the optical lattices, we then review the quantum simulation using lattice atoms in several important aspects: simulation of the Bose–Hubbard model in Chapter 13, simulation of system dynamics in Chapter 14, simulation of disordered systems in Chapter 15, and simulation of spin systems in Chapter 16.

References

[1] M. H. Anderson, J. R. Ensher, M. R. Matthews, C. E. Wieman and E. A. Cornell. *Science* **269**, 198 (1995).
[2] K. B. Davis, M.-O. Mewes, M. R. Andrews, N. J. van Druten, D. S. Durfee, D. M. Kurn and W. Ketterle. *Phys. Rev. Lett.* **75**, 3969 (1995).
[3] C. C. Bradley, C. A. Sackett, J. J. Tollett and R. G. Hulet. *Phys. Rev. Lett.* **78**, 985 (1997).
[4] C. J. Pethick and H. Smith. *Bose–Einstein Condensation in Dilute Gases*, 2nd ed. Cambridge University Press, Cambridge (2008).
[5] S. N. Bose. *Z. Phys.* **26**, 178 (1924).
[6] A. Einstein. *Sitzungsberichte der Preussischen Akademie der Wissenschaften, Physikalisch-mathematische Klasse*, 1924, p. 261.
[7] W. Ketterle, D. S. Durfee and D. M. Stamper-Kurn. Making, probing and understanding Bose–Einstein condensates, *Proceedings of the International School of Physics "Enrico Fermi"*, Course CXL. Eds. M. Inguscio, S. Stringari and C. E. Wieman. IOS Press, Amsterdam (1999).
[8] E. C. Svensson and V. F. Sears. *Progress in Low Temperature Physics*, Vol. XI. Ed. D. F. Brewer. North-Holland, Amsterdam (1987), p. 189.
[9] M. Inguscio, S. Stringari and C. E. Wieman (eds.). Bose–Einstein condensation in atomic gases, *Proceedings of the Enrico Fermi International School of Physics*. Vol. CXL. IOS Press, Amsterdam (1999).
[10] D. G. Fried, T. C. Killian, L. Willmann, D. Landhuis, S. C. Moss, D. Kleppner and T. J. Greytak. *Phys. Rev. Lett.* **81**, 3811 (1998).
[11] L. N. Cooper. *Phys. Rev.* **104**, 1189 (1956).
[12] J. Bardeen, L. N. Cooper and J. R. Schrieffer. *Phys. Rev.* **108**, 1175 (1957).
[13] E. Tiesinga, B. Verhaar and H. T. C. Stoof. *Phys. Rev. A* **47**, 4114 (1993).
[14] H. T. C. Stoof, M. Houbiers, C. A. Sackett and R. G. Hulet. *Phys. Rev. Lett.* **76**, 10 (1996).
[15] H. Feshbach. *Ann. Phys.* (N.Y.) **5**, 357 (1958).
[16] S. Inouye, M. R. Andrews, J. Stenger, H.-J. Miesner, D. M. Stamper-Kurn and W. Ketterle. *Nature* **392**, 151 (1998).

[17] P. Courteille, R. S. Freeland, D. J. Heinzen, F. A. van Abeelen and B. J. Verhaar. *Phys. Rev. Lett.* **81**, 69 (1998).
[18] J. L. Roberts, N. R. Claussen, J. P. Burke, C. H. Greene, E. A. Cornell and C. A. Wieman. *Phys. Rev. Lett.* **81**, 5109 (1998).
[19] M. Greiner, C. A. Regal and D. S. Jin. *Nature* **426**, 537 (2003).
[20] S. Jochim, M. Bartenstein, A. Altmeyer, G. Hendl, C. Chin, J. H. Denschlag and R. Grimm. *Science* **302**, 2101 (2003).
[21] M. W. Zwierlein, C. A. Stan, C. H. Schunck, S. M. F. Raupach, S. Gupta, Z. Hadzibabic and W. Ketterle. *Phys. Rev. Lett.* **91**, 250401 (2003).
[22] Q. Chen, J. Stajic, S. Tan and K. Levin. *Phys. Rep.* **412**, 1 (2005).
[23] D. Jaksch, C. Bruder, J. I. Cirac, C. W. Gardiner and P. Zoller. *Phys. Rev. Lett.* **81**, 3108 (1998).

Part I

Toward Strongly Correlated Systems

2
Atomic Structure

The study of cold atomic systems, from both experimental and theoretical perspectives, is crucially based on the understanding of atomic structures. Indeed, the abilities to confine neutral atoms in a trap, to cool them down to quantum degeneracy and to fine-tune the parameters within the system heavily rely upon the preparation and manipulation of atoms in various states. In this chapter, we discuss atomic structure of alkali-metal atoms. This choice is made based on two considerations. First, the electronic structure of an alkali-metal atom is fairly simple. There is only one valence electron while all other electrons form closed shells. Second, alkali-metal atoms are the most popular candidate for the current cold atom experiments due to their technical maturity and simplicity. However, we stress that the fundamental ideas discussed in this chapter are general and can be extended to other types of atoms.

2.1. Electronic levels of alkali-metal atoms

An alkali-metal atom with only one valence electron in the outer shell can be well described as a hydrogen-like atom with an inner core with positive charge $+|e|$ and a single valence electron with charge $-|e|$. The inner core and the valence electron interact via a Coulomb potential, which depends on their mutual distance in a good approximation. The Hamiltonian thus takes the form

$$H = -\frac{\hbar^2 \nabla_{\mathbf{r}}^2}{2m_\mu} - \frac{|e|^2}{4\pi\epsilon_0 r}, \tag{2.1}$$

where $m_\mu = Mm_e/(M + m_e)$ is the reduced mass of the nucleus with mass M and the electron with mass m_e, and ϵ_0 is the vacuum permittivity. Notice that the reduced mass can be well approximated by the electronic mass $m_\mu \approx m_e$, since m_e is at least a thousand times smaller than that of the nucleus.

This problem can be solved by noticing that the Coulomb potential acquires a spherical symmetry, which allows the resulting wavefunction to be separated into radial and angular parts

$$\Psi_{n\ell m}(\mathbf{r}) = R_{n\ell}(r)Y_{\ell m}(\hat{\mathbf{r}}), \qquad (2.2)$$

where $Y_{\ell m}(\hat{\mathbf{r}})$ are spherical harmonics labeled by the angular momentum quantum number ℓ and the magnetic quantum number m_ℓ. The radial wavefunction satisfies the so-called radial equation

$$\frac{1}{R_{n\ell}}\frac{d}{dr}\left(r^2\frac{dR_{n\ell}}{dr}\right) + \frac{2m_e r^2}{\hbar^2}\left(E + \frac{|e|^2}{4\pi\epsilon_0 r}\right) = \ell(\ell+1) \qquad (2.3)$$

with E the eigenenergy. The solution of the radial equation can be expressed as $R_{n\ell} = f_{n\ell}(r)e^{-r/na_0}$ with non-negative principal quantum number $n = 0, 1, 2, \ldots$ and $a_0 \equiv 4\pi\epsilon_0\hbar^2/m_e e^2 \approx 0.529\,\text{Å}$ the Bohr radius. The function $f_{n\ell}(r)$ is a polynomial of order $(n-1)$. The detailed forms of $f_{n\ell}$ for $n \leq 3$ are listed as below:

$$f_{10} = \frac{2}{\sqrt{a_0^3}},$$

$$f_{20} = \frac{1}{\sqrt{2a_0^3}}\left(1 - \frac{r}{2a_0}\right),$$

$$f_{21} = \frac{1}{2\sqrt{6a_0^3}}\frac{r}{a_0},$$

$$f_{30} = \frac{2}{3\sqrt{3a_0^3}}\left(1 - \frac{2r}{3a_0} + \frac{2r^2}{27a_0^2}\right),$$

$$f_{31} = \frac{4\sqrt{2}}{27\sqrt{3a_0^3}}\frac{r}{a_0}\left(1 - \frac{r}{6a_0}\right),$$

$$f_{32} = \frac{2\sqrt{2}}{81\sqrt{15a_0^3}}\frac{r^2}{a_0^2}. \qquad (2.4)$$

The eigenenergy associated with the state $\Psi_{n\ell m}$ takes the form

$$E_n = -\frac{1}{2} \frac{m_e |e|^4}{(4\pi\epsilon_0 \hbar)^2} \frac{1}{n^2}. \tag{2.5}$$

Notice that the energy is degenerate for all possible angular momenta $\ell = 0, 1, \ldots, n-1$ and magnetic quantum numbers $m \in [-\ell, \ell]$. These degeneracies will be lifted when the spin degrees of freedom of the electron and/or the nucleus are taken into account.

2.2.   Fine structure

In the Hamiltonian of Eq. (2.1), the electronic spin degrees of freedom is completely decoupled from its orbital motion around the inner core. However, if we move from the standard frame of reference where the electron orbits the nucleus, into the frame where the electron is stationary, the orbital movement of the nucleus around the electron will form a current loop, hence generating an effective magnetic field perpendicular to the orbital plane at the position of the electron. This magnetic field in turn will couple to the electronic spin, giving rise to the spin-orbit coupling term in the Hamiltonian

$$H_{\text{fs}} = \gamma_{\text{so}} \mathbf{L} \cdot \mathbf{S}, \tag{2.6}$$

where γ_{so} labels the spin-orbit coupling constant, and $\mathbf{L}$ and $\mathbf{S}$ are operators associated with the electronic orbital angular momentum and spin angular momentum, respectively.

In the presence of this spin-orbit coupling term, the Hamiltonian no longer commutes with either L_z or s_z. Instead, the total angular momentum $\mathbf{J} = \mathbf{L} + \mathbf{S}$ and its projection to the quantization axis J_z are conserved. The energy shift ΔE_J of the state with a total angular momentum J can then be calculated by noticing that

$$\mathbf{L} \cdot \mathbf{S} = \frac{1}{2} \left(\mathbf{J}^2 - \mathbf{L}^2 - \mathbf{S}^2 \right), \tag{2.7}$$

which leads to

$$\Delta E_J = \frac{\gamma_{\text{so}} \hbar^2}{2} \left[J(J+1) - \ell(\ell+1) - s(s+1) \right]. \tag{2.8}$$

For alkali-metal atoms, the inner electrons form closed shells, hence do not contribute to either spin or orbital angular momentum. The total electronic angular momentum is provided by the valence electron with

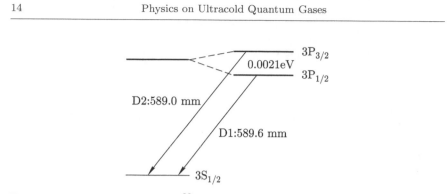

Figure 2.1.   The fine structure of ^{23}Na for the 3P states with $n = 3$ and $\ell = 1$ induced by spin-orbit coupling. This structure leads to the well-known sodium doublet around 589.3 nm in the absorption optical spectrum.

electronic spin angular momentum $s = 1/2$. As a consequence, the total angular momentum $J = \ell \pm 1/2$ and the energy shift obtained from Eq. (2.7) becomes

$$\Delta E_J = \frac{\gamma_{\text{so}}\hbar^2}{2} \begin{cases} \ell, & \text{for } J = \ell + 1/2; \\ -(\ell+1), & \text{for } J = \ell - 1/2. \end{cases} \tag{2.9}$$

Thus, the original degeneracy of the electronic levels is lifted, leading to the so-called fine structure of atomic levels. Consider the case of ^{23}Na with $n = 3$ and $\ell = 1$ as an example: the presence of spin-orbit coupling leads to two fine structure states $J = 1/2$ and $J = 3/2$ with corresponding energy shift $-\gamma_{\text{so}}\hbar^2$ and $\gamma_{\text{so}}\hbar^2/2$, respectively. This splitting causes the well-known "sodium doublet" absorption lines D1 and D2 in optical spectrum, as shown in Fig. 2.1.

2.3.   Hyperfine structure

Adopting the same semiclassical picture of electronic spin-orbit coupling, the total electronic angular momentum can also be considered as a current loop orbiting the nucleus, which usually possesses a finite magnetic dipole moment. As a result, an additional term has to be included into the Hamiltonian to describe the so-called hyperfine coupling between the electronic ($\mathbf{J}$) and nuclear ($\mathbf{I}$) angular momenta

$$H_{\text{hfs}} = \gamma_{\text{hfs}}\mathbf{I} \cdot \mathbf{J}, \tag{2.10}$$

where γ_{hfs} denotes the coupling strength. The eigenstates of this Hamiltonian are then labeled by the total angular momentum $\mathbf{F} = \mathbf{I} + \mathbf{J}$ and the energy shift is given by

$$\Delta E_F = \frac{\gamma_{\mathrm{hfs}}\hbar^2}{2}\left[F(F+1) - I(I+1) - J(J+1)\right]. \qquad (2.11)$$

Here, F and I are quantum numbers associated with $\mathbf{F}$ and $\mathbf{I}$, respectively.

For the same example of ^{23}Na, the nuclear angular momentum is $I = 3/2$. Then the $J = 1/2$ state splits into two hyperfine states with $F = 2$ and $F = 1$, respectively. The energy shifts become

$$\Delta E_F = \frac{\gamma_{\mathrm{hfs}}\hbar^2}{2}\begin{cases} 3/2, & \text{for } F = 2; \\ -5/2, & \text{for } F = 1. \end{cases} \qquad (2.12)$$

For the case of $J = 3/2$, the hyperfine splitting results in four states with $F = 0, 1, 2$ and 3, with corresponding energy shifts

$$\Delta E_F = \frac{\gamma_{\mathrm{hfs}}\hbar^2}{2}\begin{cases} 9/2, & \text{for } F = 3; \\ -3/2, & \text{for } F = 2; \\ -11/2, & \text{for } F = 1; \\ -15/2, & \text{for } F = 0. \end{cases} \qquad (2.13)$$

A schematic diagram of the hyperfine structures is shown in Fig. 2.2(a).

Another important example is the hyperfine splitting of the ground state of ^{133}Cs. In this case, the total electronic angular momentum is

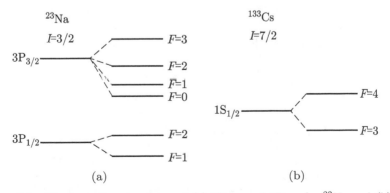

Figure 2.2. The hyperfine structures of (a) $3P_{1/2}$ and $3P_{3/2}$ for ^{23}Na and (b) the ground state $1S_{1/2}$ for ^{133}Cs.

$J = 1/2$ with $l = 0$ and $s = 1/2$, and the nuclear angular momentum is $I = 7/2$. Thus, the two hyperfine states are labeled by the total angular momentum quantum number $F = 4$ and $F = 3$. The energy difference between these two states reads

$$\delta E = \Delta E_4 - \Delta E_3 = 4\gamma_{\text{hfs}}\hbar^2. \tag{2.14}$$

The transition frequency between these two hyperfine states has been used to define the SI unit "second" since 1967.[1] A schematic diagram of the hyperfine structure of the ^{133}Cs ground state is illustrated in Fig. 2.2(b).

2.4.   Zeeman effect

The hyperfine states discussed in the previous section still acquire some degeneracy which are labeled by the magnetic quantum number m_F associated with the total angular momentum F. This degeneracy will be lifted in the presence of an external magnetic field, such that each hyperfine state will further split into a set of levels. To describe this effect, the Hamiltonian reads

$$H_{\text{Zeeman}} = \gamma_{\text{hfs}}\mathbf{I} \cdot \mathbf{J} + \gamma_e B J_z, \tag{2.15}$$

where the last term corresponds to the Zeeman coupling between the magnetic field and the electronic angular momentum with the coupling constant γ_e. Here, we have neglected the coupling of the magnetic field to the nuclear spin because the corresponding coupling strength γ_N is significantly smaller than that of an electron with $\gamma_N/\gamma_e \sim m_e/m_N$. We also assume, without any loss of generality, that the magnetic field is applied along the z-axis.

The Hamiltonian can be diagonalized by employing the raising and lowering operators $I_{\pm} = I_x \pm iI_y$ and $J_{\pm} = J_x \pm iJ_y$. Using the identity

$$\mathbf{I} \cdot \mathbf{J} = \sum_{i=x,y,z} I_i J_i = \frac{1}{2}\left(I_+ J_- + I_- J_+\right) + I_z J_z, \tag{2.16}$$

---

[1] "The second is the duration of 9,192,631,770 periods of the radiation corresponding to the transition between the two hyperfine levels of the ground state of the cesium 133 atom." SI Brochure, Sec. 2.1.1.3, Bureau International des Poids et Mesures.

it can be easily extracted that the z component of the total angular momentum is conserved, reflecting the rotational symmetry of the system with respect to the direction of the applied magnetic field, i.e., the z-axis. Thus, only states with the same value of $m_F = m_I + m_J$ are coupled by this Hamiltonian. This observation then allows us to separate the entire Hilbert space into subspaces labeled by different values of m_F and to solve for their individual eigenstates.

As an example, we consider the case of ^{23}Na with $J = 1/2$ and $I = 3/2$. This could be the ground state $1S_{1/2}$ or the excited state $3P_{1/2}$, as shown in Fig. 2.1. The magnetic quantum number for electronic angular momentum can be either $m_J = +1/2$ or $-1/2$, while it can be chosen from a set of values $m_I \in \{-3/2, -1/2, 1/2, 3/2\}$ for the nuclear spin. As a result, the value of m_F can be ± 2, ± 1 or 0. The Hamiltonian is then block diagonalized for different values of m_F.

The subspace associated with $m_F = 2$ and $m_F = -2$ contains only one single state of $|m_I = 3/2, m_J = 1/2\rangle$ and $|m_I = -3/2, m_J = -1/2\rangle$, respectively. The corresponding eigenenergy can be easily calculated as

$$E_{m_F=2} = \frac{3}{4}\gamma_{\text{hfs}} + \frac{1}{2}\gamma_e B, \tag{2.17}$$

$$E_{m_F=-2} = \frac{3}{4}\gamma_{\text{hfs}} - \frac{1}{2}\gamma_e B. \tag{2.18}$$

Notice that these two energies are both linearly dependent on the magnetic field strength.

The subspace of $m_F = 1$ contains two elements of $|m_I = 3/2, m_J = -1/2\rangle$ and $|m_I = 1/2, m_J = 1/2\rangle$. The Hamiltonian can be written within this subspace into a matrix form

$$H_{m_F=1} = \begin{pmatrix} -\frac{3}{4}\gamma_{\text{hfs}} - \frac{1}{2}\gamma_e B & \frac{\sqrt{3}}{2}\gamma_{\text{hfs}} \\ \frac{\sqrt{3}}{2}\gamma_{\text{hfs}} & -\frac{3}{4}\gamma_{\text{hfs}} + \frac{1}{2}\gamma_e B \end{pmatrix}. \tag{2.19}$$

The eigenvalues are given by the following expression

$$E_{m_F=1} = -\frac{1}{4}\gamma_{\text{hfs}} \pm \sqrt{\frac{3}{4}\gamma_{\text{hfs}}^2 + \frac{1}{4}(\gamma_{\text{hfs}} + \gamma_e B)^2}. \tag{2.20}$$

For the case of $m_F = -1$, the subspace is constructed by the two states of $|m_I = -3/2, m_J = 1/2\rangle$ and $|m_I = -1/2, m_J = -1/2\rangle$, and the eigenenergies can be obtained by simply replacing B with $-B$ in Eq. (2.20)

$$E_{m_F=-1} = -\frac{1}{4}\gamma_{\text{hfs}} \pm \sqrt{\frac{3}{4}\gamma_{\text{hfs}}^2 + \frac{1}{4}(\gamma_{\text{hfs}} - \gamma_e B)^2}. \qquad (2.21)$$

The subspace of $m_F = 0$ is spanned by $|m_I = 1/2, m_J = -1/2\rangle$ and $|m_I = -1/2, m_J = 1/2\rangle$. The Hamiltonian then takes the form

$$H_{m_F=0} = \begin{pmatrix} -\frac{1}{4}\gamma_{\text{hfs}} + \frac{1}{2}\gamma_e B & \gamma_{\text{hfs}} \\ \gamma_{\text{hfs}} & -\frac{1}{4}\gamma_{\text{hfs}} - \frac{1}{2}\gamma_e B \end{pmatrix} \qquad (2.22)$$

with eigenenergies

$$E_{m_F=0} = -\frac{1}{4}\gamma_{\text{hfs}} \pm \sqrt{\gamma_{\text{hfs}}^2 + \frac{1}{4}\gamma_e^2 B^2}. \qquad (2.23)$$

The results of Eqs. (2.17), (2.18), (2.20), (2.21) and (2.23) give all eight eigenstates of the Hamiltonian Eq. (2.15).

The magnetic field dependence of these eigenstates in the weak-field limit can be extracted by series expansion around $B = 0$. The eigenenergies for the states with $m_F = \pm 2$ are linearly dependent on B. For the $m_F = 1$ states, by expanding the corresponding results up to the second-order of B, we obtain

$$E_{m_F=1}^{(1)} = \frac{3}{4}\gamma_{\text{hfs}} + \frac{\gamma_e}{4}B + \frac{3\gamma_e^2}{32\gamma_{\text{hfs}}}B^2 + \mathcal{O}(B^3),$$

$$E_{m_F=1}^{(2)} = -\frac{5}{4}\gamma_{\text{hfs}} - \frac{\gamma_e}{4}B - \frac{3\gamma_e^2}{32\gamma_{\text{hfs}}}B^2 + \mathcal{O}(B^3). \qquad (2.24)$$

The results for the case of $m_F = -1$ can be derived from Eq. (2.24) by replacing B with $-B$. The eigenenergies associated with the two $m_F = 0$ states can be expanded similarly, leading to

$$E_{m_F=0}^{(1)} = \frac{3}{4}\gamma_{\text{hfs}} + \frac{\gamma_e^2}{8\gamma_{\text{hfs}}}B^2 + \mathcal{O}(B^3),$$

$$E_{m_F=0}^{(2)} = -\frac{5}{4}\gamma_{\text{hfs}} - \frac{\gamma_e^2}{8\gamma_{\text{hfs}}}B^2 + \mathcal{O}(B^3). \qquad (2.25)$$

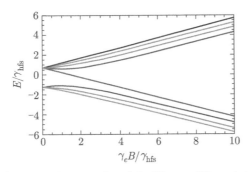

Figure 2.3. Hyperfine state eigenenergies of the $1S_{1/2}$ or $3P_{1/2}$ states of ^{23}Na atoms in the presence of an external magnetic field B. From top to bottom, the curves represent $|F = 2, m_F = 2\rangle$, $|F = 2, m_F = 1\rangle$, $|F = 2, m_F = 0\rangle$, $|F = 2, m_F = -1\rangle$, $|F = 2, m_F = -2\rangle$, $|F = 1, m_F = 1\rangle$, $|F = 1, m_F = 0\rangle$ and $|F = 1, m_F = -1\rangle$. At a zero magnetic field, the two hyperfine states with $F = 1$ and $F = 2$ have three- and five-fold degeneracies, respectively. These degeneracies are lifted at the finite magnetic field due to Zeeman splitting.

In the expressions above, the energy shifts that are linearly dependent on the magnetic field are referred to as the *linear Zeeman effect*, while the terms of B^2 are called the *quadratic Zeeman effect*. The behavior of the eigenenergies as functions of the magnetic field is shown in Fig. 2.3, where the two hyperfine states at zero magnetic field split into eight distinct levels in the presence of a magnetic field.

3
Atom-Light Interaction

The experimental achievements on cooling and trapping neutral atoms with laser light heavily rely on the understanding of atom-light interaction. Starting from the 1970s, remarkable progresses along this direction have witnessed the evolution of cold atom physics from a theoretical proposal to an astoundingly diverse field now involving multidisciplinary categories of physics, including atomic physics, nuclear physics and condensed matter physics.

One of the simplest problems involving the atom-light interaction is the coupling of a two-level atom with a single-mode light field. Although any real atom in practice acquires a complicated level structure as discussed in the previous chapter, a two-level description is valid provided that the two levels of interest are nearly resonant with the driving field, while all other levels are far detuned. In this chapter, we discuss some basics of this simplified but far-from-trivial problem, and introduce some fundamental physical properties of atom-light interacting process. To make a close connection to the most popular experimental situation, we focus on alkali-metal atoms in this chapter, and restrict our discussion to topics that are closely related to experimental schemes of laser cooling and trapping. Readers who are interested in the general theory of atom-light interaction are referred to specialized books on this subject, e.g., see Cohen-Tannoudji *et al.* [1].

3.1. Atom-light interaction Hamiltonian

As a simplified model of an alkali-metal atom, we consider a single-electron atom interacting with a light field, as described by the following

Hamiltonian

$$H = H_A + H_L - |e|\mathbf{r} \cdot \mathbf{E}. \tag{3.1}$$

Here, we have assumed that the atom-light coupling is dominated by the Coulomb interaction between the electron and the electric field component $\mathbf{E}$ of the light, with $|e|$ the electronic charge and $\mathbf{r}$ the position vector of the electron. The atomic Hamiltonian H_A can be written in the general form

$$H_A = \sum_m E_m |m\rangle\langle m|, \tag{3.2}$$

where $|m\rangle$ represents the m-th atomic eigenstate with eigenenergy E_m. In the following discussion of this chapter, we confine ourselves to the simplest case where only two atomic eigenstates are considered. Within this framework, the atomic Hamiltonian becomes

$$H_A = E_g |g\rangle\langle g| + E_e |e\rangle\langle e|, \tag{3.3}$$

where $|g\rangle$ and $|e\rangle$ denote the ground and excited states with eigenenergies E_g and E_e, respectively. Considering the complicated atomic structure discussed in the previous chapter, this two-level problem is obviously a simplified model, even for the simplest case of hydrogen atoms. However, as we will see shortly, this simplified scenario already describes many interesting aspects of atom-light interaction. The term corresponding to the light field is given in terms of the photon creation and annihilation operators

$$H_L = \sum_{\mathbf{k}} \hbar\omega_{\mathbf{k}} \left(a_{\mathbf{k}}^{\dagger} a_{\mathbf{k}} + \frac{1}{2} \right) \tag{3.4}$$

with $\omega_{\mathbf{k}}$ the frequency associated with the wave vector $\mathbf{k}$.

Using the bases of $|g\rangle$ and $|e\rangle$, the atom-light interaction term can be expressed as

$$-|e|\mathbf{r} \cdot \mathbf{E} = -|e| \sum_{m,n=(e,g)} |m\rangle\langle m|\mathbf{r} \cdot \mathbf{E}|n\rangle\langle n|$$

$$\approx \left[-|e| \sum_{m,n=(e,g)} |m\rangle\langle m|\mathbf{r}|n\rangle\langle n| \right] \cdot \mathbf{E}. \tag{3.5}$$

In the second line of the equation above, we have made an approximation that the electric component of the light is a constant within the length scale where $\langle m|\mathbf{r}|n\rangle$ makes a large contribution. In other words, we have assumed that the electric field $\mathbf{E}$ varies slowly in space, such that it remains nearly uniform over the length of atomic size. This approximation is usually referred to as the *electric dipole approximation* (EDA) or simply *dipole approximation*, indicating that the atom is considered as a point-like particle with induced electric dipole moment. In present experiments of cold atoms, lasers are usually chosen within the optical range with wavelength on the micrometer scale. Thus, the electric dipole approximation is a good starting point for atoms in their electronic ground states, in which case the atomic size is on the nanometer scale. However, for Rydberg atoms with very high principal quantum number, the extent of the atomic wavefunctions could be very large and becomes comparable to the laser wavelength. In this case, the validity of EDA becomes questionable and one has to seek other ways to better describe the system.

The electric component of the light can be expressed in terms of the photon creation and annihilation operators as [2]

$$\mathbf{E}(\mathbf{r}) = \sum_{\mathbf{k}} \mathcal{E}_{\mathbf{k}} \left(\hat{\epsilon}_{\mathbf{k}} a_{\mathbf{k}} e^{i\mathbf{k}\cdot\mathbf{r}} + \hat{\epsilon}_{\mathbf{k}}^* a_{\mathbf{k}}^\dagger e^{-i\mathbf{k}\cdot\mathbf{r}} \right), \qquad (3.6)$$

where $\mathcal{E}_{\mathbf{k}} \equiv \sqrt{\hbar\omega_{\mathbf{k}}/2\epsilon_0 \mathcal{V}}$ denotes the electric field per mode with ϵ_0 the vacuum permittivity and $\mathcal{V}$ the quantization volume. The unit vector $\hat{\epsilon}_{\mathbf{k}}$ stands for the polarization of the electric field, which is in general a complex vector for elliptically polarized lights. Notice that we have absorbed the phase factor $e^{\pm i\omega_{\mathbf{k}} t}$ associated with time into the creation and annihilation operators such that the expression Eq. (3.6) is not explicitly time dependent.

By substituting Eqs. (3.3) to (3.6) back into Eq. (3.1), the Hamiltonian becomes

$$H = \sum_{\mathbf{k}} \hbar\omega_{\mathbf{k}} a_{\mathbf{k}}^\dagger a_{\mathbf{k}} + E_g |g\rangle\langle g| + E_e |e\rangle\langle e|$$
$$+ \hbar \sum_{\mathbf{k}} (|e\rangle\langle g| + |g\rangle\langle e|) \left(\Omega_{\mathbf{k}} a_{\mathbf{k}} e^{i\mathbf{k}\cdot\mathbf{r}} + \Omega_{\mathbf{k}}^* a_{\mathbf{k}}^\dagger e^{-i\mathbf{k}\cdot\mathbf{r}} \right), \qquad (3.7)$$

where we have dropped the zero-point energy $\sum_{\mathbf{k}} \hbar\omega_{\mathbf{k}}/2$ of the light field. The parameter

$$\Omega_{\mathbf{k}} \equiv \frac{-|e|\langle e|\mathbf{r}|g\rangle \cdot \hat{\epsilon}_{\mathbf{k}} \mathcal{E}_{\mathbf{k}}}{\hbar} \qquad (3.8)$$

is defined as the *vacuum Rabi frequency*. The second and third terms of Eq. (3.7) can be rewritten as

$$E_g|g\rangle\langle g| + E_e|e\rangle\langle e| = \frac{1}{2}\left(E_g + E_e\right) + \frac{1}{2}\hbar\omega_{eg}\left(|e\rangle\langle e| - |g\rangle\langle g|\right), \qquad (3.9)$$

where $\omega_{eg} = (E_e - E_g)/\hbar$ is the transition frequency between the two atomic levels. With the aid of Pauli matrices, the Hamiltonian Eq. (3.7) can be expressed as

$$H = \sum_{\mathbf{k}} \hbar\omega_{\mathbf{k}} a_{\mathbf{k}}^{\dagger} a_{\mathbf{k}} + \frac{1}{2}\hbar\omega_{eg}\sigma_z$$

$$+ \hbar \sum_{\mathbf{k}} (\sigma_+ + \sigma_-)\left(\Omega_{\mathbf{k}} a_{\mathbf{k}} e^{i\mathbf{k}\cdot\mathbf{r}} + \Omega_{\mathbf{k}}^* a_{\mathbf{k}}^{\dagger} e^{-i\mathbf{k}\cdot\mathbf{r}}\right), \qquad (3.10)$$

where $\sigma_{\pm} = \sigma_x \pm i\sigma_y$ are the raising and lowering operators, and the constant $(E_g + E_e)/2$ has been dropped.

The interaction term in the resulting Hamiltonian Eq. (3.10) contains four parts. The term associated with $\sigma_+ a_{\mathbf{k}}$ corresponds to the process by which an atom in the ground state absorbs a photon and is pumped up to the excited state, as illustrated in Fig. 3.1(a). The term $\sigma_- a_{\mathbf{k}}^{\dagger}$ describes the opposite process by which an atom in the excited state emits a photon and ends up in the ground state, as shown in Fig. 3.1(b). If the energy of the absorbed or emitted photon is equivalent to the energy splitting of the two atomic levels, energy is conserved in these two processes. The third term associated with $\sigma_+ a_{\mathbf{k}}^{\dagger}$ corresponds to the process through which an atom in the ground state emits a photon and is stimulated to the excited state,

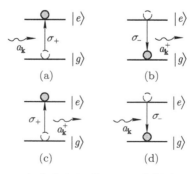

Figure 3.1.   Four processes included in the atom-light interaction Hamiltonian Eq. (3.10). In the rotating-wave approximation, the processes correspond to (c) and (d) are neglected.

while the fourth term $\sigma_- a_{\mathbf{k}}$ describes the process by which an atom in the excited state absorbs a photon and comes down to the ground state. These two processes are depicted in Figs. 3.1(c) and 3.1(d), respectively.[1] In the rotating-wave approximation (RWA), the third and the fourth terms are neglected, leading to the simplified Hamiltonian

$$H = \sum_{\mathbf{k}} \hbar \omega_{\mathbf{k}} a_{\mathbf{k}}^{\dagger} a_{\mathbf{k}} + \frac{1}{2} \hbar \omega_{eg} \sigma_z + \hbar \sum_{\mathbf{k}} \left(\Omega_{\mathbf{k}} \sigma_+ a_{\mathbf{k}} e^{i \mathbf{k} \cdot \mathbf{r}} + \Omega_{\mathbf{k}}^* \sigma_- a_{\mathbf{k}}^{\dagger} e^{-i \mathbf{k} \cdot \mathbf{r}} \right).$$

$$(3.11)$$

This Hamiltonian is the starting point of the remaining discussion of this chapter.

3.2. Spontaneous emission

In 1916, Albert Einstein categorized the atom-light coupling processes into three types [3], including

- *Spontaneous Emission*: a process by which an atom initially in the excited state undergoes a transition to the ground state and emits a photon.
- *Stimulated Emission*: a process by which an atom initially in an excited state interacting with a light field drops to the ground state, transferring the residual energy to the light field.
- *Stimulated Absorption*: a process by which an atom initially in the ground state absorbs a photon from a light field and jumps to the excited state.

Each process is associated with a so-called Einstein coefficient, which measures the probability of the occurrence of that particular process. In this section, we investigate the spontaneous emission process with the aid of the atom-light interacting Hamiltonian of Eq. (3.11). The discussion on the two stimulated processes will be found in the next section.

To begin with, we consider an atom at the origin, and separate the Hamiltonian Eq. (3.11) into the non-interacting and the interaction parts

---

[1] Notice that the last two processes are not conserving energy, even in the case where the light is on resonance with the two-level atom, but respectively correspond to a gaining and losing of a finite amount of energy. However, the gaining and losing of energy terms are always present in pair in the Hamiltonian, describing a virtual process of exchanging energy with the vacuum field, hence are perfectly allowed.

$H = H_0 + H_{\text{int}}$ with

$$H_0 = \sum_{\mathbf{k}} \hbar \omega_{\mathbf{k}} a_{\mathbf{k}}^{\dagger} a_{\mathbf{k}} + \frac{1}{2} \hbar \omega_{eg} \sigma_z,$$

$$H_{\text{int}} = \hbar \sum_{\mathbf{k}} \left(\Omega_{\mathbf{k}} \sigma_+ a_{\mathbf{k}} + \Omega_{\mathbf{k}}^* \sigma_- a_{\mathbf{k}}^{\dagger} \right). \qquad (3.12)$$

It is convenient to work in the interaction picture, where the Hamiltonian is given by

$$H^{\text{I}} = e^{iH_0 t/\hbar} H_{\text{int}} e^{-iH_0 t/\hbar}. \qquad (3.13)$$

Since H_0 and H_{int} do not commute, we employ the following identity

$$e^{\alpha \hat{A}} \hat{B} e^{-\alpha \hat{A}} = \hat{B} + \alpha \left[\hat{A}, \hat{B} \right] + \frac{\alpha^2}{2!} \left[\hat{A}, \left[\hat{A}, \hat{B} \right] \right] + \cdots, \qquad (3.14)$$

where $\hat{A}$ and $\hat{B}$ are operators, and $[\cdot, \cdot]$ is the commutator. It is then straightforward to demonstrate that

$$e^{i\omega_{\mathbf{k}} a_{\mathbf{k}}^{\dagger} a_{\mathbf{k}} t} a_{\mathbf{k}} e^{-i\omega_{\mathbf{k}} a_{\mathbf{k}}^{\dagger} a_{\mathbf{k}} t} = a_{\mathbf{k}} e^{-i\omega_{\mathbf{k}} t},$$

$$e^{i\omega_{eg} \sigma_z t/2} \sigma_+ e^{-i\omega_{eg} \sigma_z t/2} = \sigma_+ e^{i\omega_{eg} t}. \qquad (3.15)$$

By substituting Eq. (3.15) into the interaction-picture Hamiltonian Eq. (3.13), we obtain

$$H^{\text{I}} = \hbar \sum_{\mathbf{k}} \left[\Omega_{\mathbf{k}} \sigma_+ a_{\mathbf{k}} e^{i(\omega_{eg} - \omega_{\mathbf{k}})t} + \text{H.C.} \right], \qquad (3.16)$$

where H.C. stands for the Hermitian conjugate.

   To analyze the process of spontaneous decay, we assume that at time $t = 0$ the atom is in the excited state $|e\rangle$ and the light field mode is in the vacuum state. This state can be written as $|e, 0\rangle$, where 0 represents that there is no photon in the space. As time evolves, the atom has the probability to drop down to the ground state $|g\rangle$ and emit a photon with wavevector $\mathbf{k}$, leading to a new state of $|g, 1_{\mathbf{k}}\rangle$, where 1 indicates that there is one photon in the space. Thus, the time evolution of the quantum state of the system takes the form

$$|\psi(t)\rangle = c_e(t)|e, 0\rangle + \sum_{\mathbf{k}} c_{g,\mathbf{k}}(t)|g, 1_{\mathbf{k}}\rangle \qquad (3.17)$$

with initial conditions $c_e(t = 0) = 1$ and $c_{g,\mathbf{k}}(t = 0) = 0$.

By substituting the wavefunction Eq. (3.17) into the Schrödinger equation

$$\frac{d|\psi(t)\rangle}{dt} = -\frac{i}{\hbar}H^{\mathrm{I}}|\psi(t)\rangle \qquad (3.18)$$

and matching terms properly, we obtain the equation of motion for the coefficients $c_e(t)$ and $c_{g,\mathbf{k}}(t)$

$$\frac{dc_e(t)}{dt} = -i\sum_{\mathbf{k}} c_{g,\mathbf{k}}(t)\Omega_{\mathbf{k}}e^{-i(\omega_{\mathbf{k}}-\omega_{eg})t}, \qquad (3.19a)$$

$$\frac{dc_{g,\mathbf{k}}(t)}{dt} = -ic_e(t)\Omega_{\mathbf{k}}^*e^{i(\omega_{\mathbf{k}}-\omega_{eg})t}. \qquad (3.19b)$$

The differential equations for $c_e(t)$ and $c_{g,\mathbf{k}}(t)$ listed above should be self-consistently solved. In fact, we can first integrate Eq. (3.19b) and then substitute it back into Eq. (3.19a) to eliminate $c_{g,\mathbf{k}}(t)$. This procedure leads to the results

$$\frac{dc_e(t)}{dt} = -\sum_{\mathbf{k}} |\Omega_{\mathbf{k}}|^2 \int_0^t dt' e^{-i(\omega_{\mathbf{k}}-\omega_{eg})(t-t')}c_e(t'), \qquad (3.20a)$$

$$c_{g,\mathbf{k}}(t) = -i\Omega_{\mathbf{k}}^* \int_0^t dt' e^{i(\omega_{\mathbf{k}}-\omega_{eg})t'}c_e(t'). \qquad (3.20b)$$

Notice that we have employed the initial condition $c_{g,\mathbf{k}}(t = 0) = 0$ to obtain Eq. (3.20b). The integral-differential Eq. (3.20a) for $c_e(t)$ describes the spontaneous decay of the excited state. To obtain an explicit expression for the decay rate, we next evaluate the summation over momentum and the integration over time.

In order to count the number of modes of light field, we consider a complete set of linearly polarized traveling waves in a cube with each side L. The x-, y- and z-components of $\mathbf{k}$ are quantized as $k_{i=x,y,z} = 2\pi n_i/L$, where n_i are non-negative integers. Besides, since there are two independent polarization vectors that are perpendicular to the wavevector $\mathbf{k}$, the total number of modes within a given volume in momentum space is $dn = 2 \times (L/2\pi)^3 d^3\mathbf{k}$. As a result, the summation over momentum can be transformed into an integral form

$$\sum_{\mathbf{k}} \rightarrow 2 \times \frac{L^3}{(2\pi)^3} \int_0^{2\pi} d\phi \int_0^{\pi} \sin\theta d\theta \int_0^{\infty} k^2 dk. \qquad (3.21)$$

In addition, since the lights are linearly polarized, we can assume, without loss of generality, that both the polarization unit vector $\hat{\epsilon}_{\mathbf{k}}$ and the electric dipole moment $\mathbf{p}_{eg} \equiv |e|\langle e|\mathbf{r}|g\rangle$ are real. As a result, using the definition given in Eq. (3.8), the square of the vacuum Rabi frequency $\Omega_{\mathbf{k}}$ can be written as

$$|\Omega_{\mathbf{k}}|^2 = \frac{\omega_{\mathbf{k}}}{2\hbar\epsilon_0 \mathcal{V}} |\mathbf{p}_{eg}|^2 \cos^2(\theta), \qquad (3.22)$$

where θ represents the angle between the two real vectors $\mathbf{p}_{eg}$ and $\hat{\epsilon}_{\mathbf{k}}$.

Substituting Eqs. (3.21) and (3.22) into Eq. (3.20a), the integral-differential equation for $c_e(t)$ can be rewritten as

$$\frac{dc_e(t)}{dt} = -\frac{4|\mathbf{p}_{eg}|^2}{(2\pi^2)6\hbar\epsilon_0 c^3} \int_0^{\infty} d\omega_{\mathbf{k}} \omega_{\mathbf{k}}^3 \int_0^t dt' e^{-i(\omega_{\mathbf{k}}-\omega_{eg})(t-t')} c_e(t'). \qquad (3.23)$$

Here, we have used the relation $\omega_{\mathbf{k}} = ck$ with c being the speed of light in vacuum, and carried out the integrations over the polar angle θ and azimuth angle ϕ explicitly. Next, we interchange the sequence of the integrations over $\omega_{\mathbf{k}}$ and t', and notice that the major contribution to the integral over $\omega_{\mathbf{k}}$ is from a small region centered around $\omega_{\mathbf{k}} \sim \omega_{eg}$ due to the fast oscillation of the exponential term. Thus, we have

$$\int_0^{\infty} d\omega_{\mathbf{k}} \omega_{\mathbf{k}}^3 e^{-i(\omega_{\mathbf{k}}-\omega_{eg})(t-t')} \approx \omega_{eg}^3 \int_{-\infty}^{\infty} d\omega_{\mathbf{k}} e^{-i(\omega_{\mathbf{k}}-\omega_{eg})(t-t')}$$

$$= 2\pi\omega_{eg}^3 \delta(t - t'). \qquad (3.24)$$

In the first line, we also extend the lower bound of the integral from 0 to $-\infty$, since the contribution from the negative frequency part is negligible. This procedure is known as the *Weisskopf–Wigner approximation*. Thus, we can rewrite Eq. (3.23) as

$$\frac{dc_e(t)}{dt} = -\frac{\Gamma}{2} c_e(t). \qquad (3.25)$$

Obviously, the solution of this differential equation with initial condition $c_e(t = 0) = 1$ is an exponential decay $c_e(t) = e^{-\Gamma t/2}$ with the decay constant

$$\Gamma = \frac{1}{4\pi\epsilon_0} \frac{4\omega_{eg}^3 |\mathbf{p}_{eg}|^2}{3\hbar c^3}. \tag{3.26}$$

Notice that the lifetime of the atom in the excited state is $\tau = 1/\Gamma$.

3.3. Stimulated absorption and emission

In comparison to the process of spontaneous emission discussed in the previous section, the key difference of stimulated absorption and emission processes is that there is an existing light field, and the atom can only exchange photons with this light field. In other words, the photon absorbed or emitted by the atom must have the same frequency, wavevector and polarization as the existing light field. As the simplest case, we consider a two-level atom interacting with a single-mode laser light, where the atom-light interaction Hamiltonian Eq. (3.11) can be reduced to the following form

$$H = \hbar\omega_\ell a_{\mathbf{k}}^\dagger a_{\mathbf{k}} + \frac{1}{2}\hbar\omega_{eg}\sigma_z + \hbar\left(\Omega_{\mathbf{k}}\sigma_+ a_{\mathbf{k}} + \Omega_{\mathbf{k}}\sigma_- a_{\mathbf{k}}^\dagger\right). \tag{3.27}$$

Here, the single-mode laser light is characterized by a wavevector $\mathbf{k}$ and frequency ω_ℓ, and $\Omega_{\mathbf{k}}$ is assumed to be real. This Hamiltonian is known as the *Jaynes–Cummings model* [4], which is of great interest in atomic physics, quantum optics and solid-state quantum information circuits.

3.3.1. *Rabi oscillation*

The first step toward the understanding of this two-level problem is to fix the number of photons of the light field and take the following wavefunction ansatz

$$|\psi(t)\rangle = c_e(t)|e, N-1\rangle + c_g(t)|g, N\rangle, \tag{3.28}$$

where N is the total number of photons when the atom is in the ground state. Comparing to the state ansatz Eq. (3.17) of spontaneous emission, the N photons here all belong to the same mode as the existing laser light. This treatment is called the *semiclassical approximation*, which assumes that the light is a classical field and completely neglects its quantum fluctuation. The position-dependent light-field intensity is then proportional to the local number of photons N.

Substituting $|\psi(t)\rangle$ into the Schrödinger equation, we obtain the equations for $c_e(t)$ and $c_g(t)$

$$\frac{dc_e(t)}{dt} = -i(\omega_{eg} - \omega_\ell)c_e(t) - i\frac{\Omega}{2}c_g(t), \qquad (3.29a)$$

$$\frac{dc_g(t)}{dt} = -i\frac{\Omega^*}{2}c_e(t). \qquad (3.29b)$$

To derive the equations above, we have shifted the zero point of energy to $N\hbar\omega_\ell - \hbar\omega_{eg}/2$. We have also defined the Rabi frequency of the light as $\Omega \equiv 2\sqrt{N}\Omega_{\mathbf{k}}$. Considering the definition of $\Omega_{\mathbf{k}}$ in Eq. (3.8) and the quantization of electric field Eq. (3.6), the Rabi frequency Ω can be looked as

$$\Omega = \frac{-|e|\langle e|\mathbf{r}|g\rangle \cdot \mathbf{E}}{\hbar}, \qquad (3.30)$$

where the creation and annihilation operators $a_{\mathbf{k}}^\dagger \approx a_{\mathbf{k}} \approx \sqrt{N}$ in the semiclassical approximation. Hence, the Rabi frequency is proportional to the square root of the laser intensity $I \propto |\mathbf{E}|^2$.

This set of equations can be solved by replacing $c_e(t)$ in Eq. (3.29a) with the aid of Eq. (3.29b), and by replacing $c_g(t)$ in Eq. (3.29b) with the aid of Eq. (3.29a). This procedure leads to

$$\frac{d^2c_g(t)}{dt^2} - i\delta\frac{dc_g(t)}{dt} + \frac{\Omega^2}{4}c_g(t) = 0, \qquad (3.31a)$$

$$\frac{d^2c_e(t)}{dt^2} + i\delta\frac{dc_e(t)}{dt} + \frac{\Omega^2}{4}c_e(t) = 0. \qquad (3.31b)$$

Here, we consider without loss of generality, an atom at the origin interacting with a linearly polarized light field, such that the Rabi frequency is real. Besides, we also define a quantity called *detuning* as $\delta \equiv \omega_\ell - \omega_{eg}$ to represent the difference between the frequency of the light field and the atomic resonance frequency. The general solutions of Eq. (3.31) take the form

$$c_g(t) = a_1 e^{i\omega_1 t} + a_2 e^{i\omega_2 t}, \quad \text{with } \omega_{1,2} = \frac{\delta \pm \sqrt{\delta^2 + \Omega^2}}{2}; \qquad (3.32a)$$

$$c_e(t) = b_1 e^{i\nu_1 t} + b_2 e^{i\nu_2 t}, \quad \text{with } \nu_{1,2} = \frac{-\delta \pm \sqrt{\delta^2 + \Omega^2}}{2}. \qquad (3.32b)$$

The coefficients $a_{1,2}$ and $b_{1,2}$ must be determined by initial conditions of c_g and c_e, together with the normalization condition $|c_g(t)|^2 + |c_e(t)|^2 = 1$.[2]
    The general solutions of $c_g(t)$ and $c_e(t)$ obviously represent an oscillatory behavior of the probability for the atom to be found in the ground or the excited state. As a particular example, we consider the case where the atom is in the ground state at $t = 0$. The initial conditions hence are given by $c_g(t = 0) = 1$ and $c_e(t = 0) = 0$. Therefore, the time evolutions of $c_g(t)$ and $c_e(t)$ for $t > 0$ take the following forms

$$c_g(t) = \left(\cos \frac{\Omega' t}{2} - i \frac{\delta}{\Omega'} \sin \frac{\Omega' t}{2} \right) e^{i\delta t/2}, \qquad (3.33a)$$

$$c_e(t) = -i \frac{\Omega}{\Omega'} \sin \frac{\Omega' t}{2} e^{-i\delta t/2}, \qquad (3.33b)$$

where $\Omega' \equiv \sqrt{\Omega^2 + \delta^2}$. It is straightforward to see that the probabilities for finding the atom in the ground or the excited states oscillate at frequency Ω'

$$|c_g(t)|^2 = \frac{\delta^2}{\Omega^2 + \delta^2} + \frac{\Omega^2}{\Omega^2 + \delta^2} \cos^2 \frac{\Omega' t}{2}, \qquad (3.34a)$$

$$|c_e(t)|^2 = \frac{\Omega^2}{\Omega^2 + \delta^2} \sin^2 \frac{\Omega' t}{2}. \qquad (3.34b)$$

    For the resonant case with $\delta = 0$, the oscillation amplitude is unity, indicating that the atom will be transferred back and forth between the ground and excited states. The oscillation frequency is the Rabi frequency Ω, which is named after the 1937 seminal paper by Isidor Rabi, who analyzed the problem of a magnetic dipole undergoing precessions of an applied radio-frequency magnetic field [5]. By increasing the detuning $|\delta|$ from zero, the oscillation amplitude will be reduced from unity and the frequency will be increased. The behavior of $|c_e(t)|^2$ for different values of $|\delta|/\Omega$ are shown in Fig. 3.2.

---

[2]Notice that the four free parameters in the general solutions Eq. (3.32) cannot be fully determined by the two initial conditions plus the normalization condition. This is because we implicitly increase the order of the differential equations by writing down Eq. (3.31), and hence introduce an artificial free parameter. Thus, we cannot solve the two equations in Eq. (3.31) simultaneously. Instead, we should solve one equation in Eq. (3.31) and another one in Eq. (3.29).

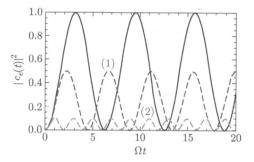

Figure 3.2. Time evolution of the probability for the atom on the excited state $|c_e(t)|^2$. The three curves correspond to $|\delta/\Omega| = 0$ (solid curve), 1 (dashed curve (1)), and 3 (dashed curve (2)). Notice that for larger detuning, the oscillation acquires an enhanced frequency and a reduced amplitude.

3.3.2. Energy shifts

In the presence of the atom-light interaction, the atomic energy levels E_g and E_e are no longer the eigenvalues of the full Hamiltonian. To solve for the energy shifts, we can write Eq. (3.29) into a matrix form

$$i\hbar\frac{d}{dt}\begin{pmatrix} c_e(t) \\ c_g(t) \end{pmatrix} = \hbar \begin{pmatrix} -\delta & \Omega/2 \\ \Omega/2 & 0 \end{pmatrix} \begin{pmatrix} c_e(t) \\ c_g(t) \end{pmatrix} = H' \begin{pmatrix} c_e(t) \\ c_g(t) \end{pmatrix}. \qquad (3.35)$$

The two eigenvalues of the matrix H' are given by

$$\lambda_{1,2} = \frac{\hbar}{2}\left(-\delta \pm \Omega'\right). \qquad (3.36)$$

The eigenvectors associated with the two eigenvalues are usually referred to as the *dressed states* of the atom, which can be written as

$$|\psi_{1,2}\rangle = \mathcal{N}\left(-\frac{\delta \mp \Omega'}{\Omega}|e\rangle + |g\rangle\right), \qquad (3.37)$$

where $\mathcal{N}$ is the normalization constant.

The one-to-one correspondence between $\lambda_{1,2}$ and the shifted energies $E'_{e,g}$ depends on the sign of detuning. Indeed, if the detuning is negative with $\delta \equiv \omega_\ell - \omega_{eg} < 0$, then λ_1 with a plus sign in the bracket corresponds to E'_e, as one can see from the limiting case of $\Omega \to 0$. In this situation, we usually say that the light is *red-detuned* from the atomic resonance since the light has a lower frequency. On the other hand, if the detuning is positive with $\delta > 0$, one can easily see that λ_1 corresponds to E'_g and the light is

said to be *blue-detuned*. In summary, the shifted energies of the ground and
excited states become

$$E'_g = \lambda_1 \text{ (red-detuned)}, \quad \lambda_2 \text{ (blue-detuned)};$$

$$E'_e = \lambda_2 \text{ (red-detuned)}, \quad \lambda_1 \text{ (blue-detuned)};} \tag{3.38}$$

In the limiting case where $\Omega \gg |\delta|$, the resulting energy shifts induced
by the presence of light are

$$\Delta E_g = \frac{\hbar \Omega^2}{4\delta}, \tag{3.39a}$$

$$\Delta E_e = -\frac{\hbar \Omega^2}{4\delta}. \tag{3.39b}$$

In the opposite limit of $\Omega \ll |\delta|$, the energy shifts can be written as

$$\Delta E_g = \text{sgn}(\delta) \frac{\hbar \Omega}{2}, \tag{3.40a}$$

$$\Delta E_e = -\text{sgn}(\delta) \frac{\hbar \Omega}{2}. \tag{3.40b}$$

In Fig. 3.3, we plot the energy shifts as functions of δ/Ω. Notice that for
the red-detuned case, the energy shift of the ground state ΔE_g is always
negative and is monotonically dependent on the Rabi frequency Ω. Since Ω^2
is proportional to the light intensity, the energy of an atom in the ground
state would be spatially dependent in a nonuniform light field. As a result,
the atom prefers to stay at the position of the lowest energy, where the
light intensity is the strongest to induce a maximal energy shift. If an atom
is deviated from this optimal spot, it will experience a force pushing the

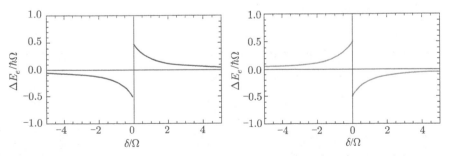

Figure 3.3.   The energy shifts of the ground (left) and excited (right) states of the atom
induced by the atom-light interaction.

atom back. This force induced by the gradient of energy is called the *dipole force* and will be discussed in more detail in Sec. 3.5.

3.4. The optical Bloch equations

In the previous two sections, we have analyzed the effect of atom-light interaction in a two-level atom. However, in both cases, we have confined ourselves within a subspace of fixed number of photons. While the photon number is zero when investigating spontaneous emission, it is N for the case of stimulated processes. It is apparent that a unified theory that can describe all three absorption and emission processes must relax this constraint, and incorporate the entire Hilbert space with an arbitrary number of photons having different frequencies, wave-vectors and polarizations. In this general situation, the method we have used successfully in the previous two sections would not work, since the ansatz state now contains an infinite number of terms, each corresponding to a different configuration of photons. At the same time, we notice that in most practical cases, the troublesome detailed information of photons are not crucial and it is the state of atom that really matters. Therefore, we would like to seek a practical way to study the behavior of atoms, while leaving the exact state of photons unknown. This goal can be achieved by the assistance of a density matrix.

3.4.1. *Density matrix*

Density matrix is the matrix representation of the density operator of a given mixed state

$$\rho \equiv \sum_j p_j |\psi_j\rangle\langle\psi_j|, \tag{3.41}$$

where p_j is the probability for the system to be found in state $|\psi_j\rangle$. In an orthonormal basis of the entire Hilbert space $\{|\phi_n\rangle\}$, the state $|\psi_j\rangle$ can be expanded as

$$|\psi_j\rangle = \sum_n c_{jn}|\phi_n\rangle. \tag{3.42}$$

Thus, the density matrix takes the form

$$\rho_{mn} = \sum_j p_j c_{jm} c_{jn}^*. \tag{3.43}$$

It is straightforward to verify that a density matrix has the following properties

$$\begin{cases} \mathrm{Tr}(\rho) = 1, \\ \rho^2 = \rho, \quad \text{for a pure state,} \\ \rho^2 \neq \rho, \quad \text{for a mixed state.} \end{cases}$$

The expectation value of an operator $\mathcal{O}$ for this mixed state can be written as

$$\langle \mathcal{O} \rangle = \sum_j p_j \langle \psi_j | \mathcal{O} | \psi_j \rangle = \sum_j p_j \left(\sum_m \langle \phi_m | c_{jm}^* \right) \mathcal{O} \left(\sum_n c_{jn} | \phi_n \rangle \right)$$

$$= \sum_{m,n} \sum_j p_j c_{jm}^* c_{jn} \langle \phi_m | \mathcal{O} | \phi_n \rangle = \sum_{m,n} \rho_{nm} \mathcal{O}_{mn}$$

$$= \sum_n (\rho \mathcal{O})_{nn} = \mathrm{Tr}\,(\rho \mathcal{O}) . \tag{3.44}$$

The equation of motion for a density matrix can be derived from the Schrödinger equation by noticing that

$$\frac{d\rho}{dt} = \sum_j p_j \left(\frac{d|\psi_j\rangle}{dt} \langle \psi_j | + |\psi_j\rangle \frac{d\langle \psi_j|}{dt} \right)$$

$$= -\frac{i}{\hbar} \left(H \sum_j p_j |\psi_j\rangle\langle \psi_j | - \sum_j p_j |\psi_j\rangle\langle \psi_j | H \right)$$

$$= -\frac{i}{\hbar} [H, \rho] . \tag{3.45}$$

Notice that this equation only holds when the density operator is taken in the Schrödinger picture, although it has a similar form of the Heisenberg equation of motion in the Heisenberg picture with an apparent opposite sign on the right-hand side. The equation of motion (3.45) is called the *von Neumann equation* or the *quantum Liouville equation*.

For the specific problem of a two-level atom interacting with a single-mode light field, the equation of motion for the atomic density matrix can be derived by independently incorporating the stimulated processes and spontaneous emission. In particular, the atomic density matrix for a pure state can be written as

$$\rho \equiv \begin{pmatrix} \rho_{ee} & \rho_{eg} \\ \rho_{ge} & \rho_{gg} \end{pmatrix} = \begin{pmatrix} c_e c_e^* & c_e c_g^* \\ c_g c_e^* & c_g c_g^* \end{pmatrix}. \tag{3.46}$$

The stimulated processes can be described by substituting the equations of motion (3.29) for $c_e(t)$ and $c_g(t)$, leading to

$$\frac{d\rho_{ee}}{dt} = \frac{i}{2}\left(\Omega^* \rho_{eg} - \Omega \rho_{ge}\right),$$

$$\frac{d\rho_{gg}}{dt} = \frac{i}{2}\left(-\Omega^* \rho_{eg} + \Omega \rho_{ge}\right),$$

$$\frac{d\rho_{eg}}{dt} = i\delta\rho_{eg} + \frac{i}{2}\Omega\left(\rho_{ee} - \rho_{gg}\right),$$

$$\frac{d\rho_{ge}}{dt} = -i\delta\rho_{ge} + \frac{i}{2}\Omega^*\left(-\rho_{ee} + \rho_{gg}\right). \tag{3.47}$$

To incorporate the effect of spontaneous emission, an additional exponential decay has to be included for $c_e(t)$ with a constant decay rate $\Gamma/2$ given by Eq. (3.26). As a result, we obtain the final equations of motion for the atomic density matrix

$$\frac{d\rho_{ee}}{dt} = -\Gamma\rho_{ee} + \frac{i}{2}\left(\Omega^* \rho_{eg} - \Omega \rho_{ge}\right),$$

$$\frac{d\rho_{gg}}{dt} = \Gamma\rho_{ee} + \frac{i}{2}\left(-\Omega^* \rho_{eg} + \Omega \rho_{ge}\right),$$

$$\frac{d\rho_{eg}}{dt} = \left(-\frac{\Gamma}{2} + i\delta\right)\rho_{eg} + \frac{i}{2}\Omega\left(\rho_{ee} - \rho_{gg}\right),$$

$$\frac{d\rho_{ge}}{dt} = \left(-\frac{\Gamma}{2} - i\delta\right)\rho_{ge} + \frac{i}{2}\Omega^*\left(-\rho_{ee} + \rho_{gg}\right). \tag{3.48}$$

These equations are called the *optical Bloch equations* (OBEs) [6], in analogy to the Bloch equations describing the time evolution of magnetization in the context of nuclear magnetic resonance, electron spin resonance, as well as magnetic resonance imaging. Notice that the time derivatives satisfy $d(\rho_{ee} + \rho_{gg})/dt = 0$, which is consistent with the requirement that the total population of the two levels $\rho_{ee} + \rho_{gg} = 1$ is conserved.

3.4.2. *Steady-state solutions*

In this section, we discuss the steady-state solutions of the OBEs, which can be obtained by setting all time derivatives to zero. Considering the fact that ρ_{ee} and ρ_{gg} are real, and that ρ_{eg} and ρ_{ge} are complex conjugates, there are five coefficients in the density matrix in total. These coefficients can be solved from the four OBEs and the normalization condition $\rho_{ee} + \rho_{gg} = 1$. The solutions can be expressed as

$$\rho_{ee} = \frac{1-p}{2},$$

$$\rho_{gg} = \frac{1+p}{2},$$

$$\rho_{eg} = \frac{\Omega}{2\delta + i\Gamma}p,$$

$$\rho_{ge} = \frac{\Omega^*}{2\delta - i\Gamma}p. \tag{3.49}$$

Here, $p \equiv \rho_{gg} - \rho_{ee}$ is the population difference, which is given by

$$p = \frac{1}{1+s} \tag{3.50}$$

with the saturation parameter

$$s = \frac{2|\Omega|^2/\Gamma^2}{1 + 4\delta^2/\Gamma^2}. \tag{3.51}$$

For the case of a small saturation parameter, $s \ll 1$, the population of the excited state is negligible with $p \approx 1$. On the other hand, if the saturation parameter is large, the atom is equally populated between the ground and the excited states with $p \to 0$. In general cases, the population of the excited state reads

$$\rho_{ee} = \frac{s}{2(1+s)}, \tag{3.52}$$

which is bounded within the region $0 \le \rho_{ee} \le 1/2$.

3.5. Light forces on atoms

Based on the knowledge of atom-light interaction, we are at the stage to understand the force exerted by the laser light on a neutral atom. The force

F is defined as the expectation value of the force operator

$$\mathbf{F} = \langle \mathcal{F} \rangle = \frac{d}{dt} \langle \mathbf{p} \rangle, \tag{3.53}$$

where **p** is the momentum operator. With the aid of the Heisenberg equation

$$\frac{d \langle \mathbf{p} \rangle}{dt} = \frac{i}{\hbar} \langle [H, \mathbf{p}] \rangle, \tag{3.54}$$

the force on an atom is then given by

$$\mathbf{F} = -\langle \nabla H \rangle. \tag{3.55}$$

Here, we have exploited the spatial representation for the momentum operator $\mathbf{p} = -i\hbar\nabla$. Equation (3.55) is clearly the quantum mechanical analog of the classical mechanical relation stating that the force is the negative gradient of the potential.

The expectation value of ∇H can be evaluated by substituting the Hamiltonian Eq. (3.27) within the semiclassical approximation, and using the atomic density matrix Eq. (3.46) and the relation $\langle \mathcal{O} \rangle = \text{Tr}(\rho\mathcal{O})$ of Eq. (3.44). This procedure leads to

$$\mathbf{F} = -\text{Tr}\,(\rho \nabla H)$$

$$= -\frac{\hbar}{2}\text{Tr}\begin{pmatrix} \rho_{ee} & \rho_{eg} \\ \rho_{ge} & \rho_{gg} \end{pmatrix}\begin{pmatrix} 0 & \nabla\Omega \\ \nabla\Omega^* & 0 \end{pmatrix}$$

$$= -\frac{\hbar}{2}\left(\rho_{eg}\nabla\Omega^* + \rho_{ge}\nabla\Omega\right). \tag{3.56}$$

For our convenience, we separate the spatial dependence of Ω into the magnitude and phase parts $\Omega(\mathbf{r}) \equiv \Omega_0(\mathbf{r})e^{i\theta(\mathbf{r})t}$, where $\Omega_0(\mathbf{r})$ and $\theta(\mathbf{r})$ are real functions. By substituting the steady-state solutions Eq. (3.49) of the OBEs into Eq. (3.56), we get

$$\mathbf{F} = -\frac{\hbar}{2}\frac{s}{1+s}\left(\frac{2\delta\nabla\Omega_0}{\Omega_0} - \Gamma\nabla\theta\right). \tag{3.57}$$

Notice that the first term is proportional to the detuning and the spatial derivative of the Rabi frequency magnitude, while the second term is proportional to the spontaneous decay rate and the phase derivative.

Obviously, the second term in the light force is rooted in the spontaneous emission process. To see this more transparently, we consider a two-level atom interacting with a linearly polarized traveling wave as an example, where the Rabi frequency can be written as

$$\Omega = \Omega_\ell e^{i\mathbf{k}\cdot\mathbf{r}}. \tag{3.58}$$

Here, $\Omega_\ell = -|e|\langle e|\mathbf{r}|g\rangle \cdot \mathbf{E}_0/\hbar$ is a real constant with $\mathbf{E}_0$ the electric field amplitude. Thus, the spatial derivative of Ω_0 vanishes and the gradient of the phase $\nabla\theta = \mathbf{k}$. As a result, the light force Eq. (3.57) in this case reduces to

$$\begin{aligned}
\mathbf{F}_{\text{rad}} &= \frac{\hbar\mathbf{k}\Gamma}{2}\frac{s}{1+s} = \hbar\mathbf{k}\Gamma\rho_{ee} \\
&= \frac{\hbar\mathbf{k}\Gamma\Omega_\ell^2}{\Gamma^2 + 4\delta^2 + 2\Omega_\ell^2},
\end{aligned} \tag{3.59}$$

where the result of Eq. (3.52) for ρ_{ee} is employed. Notice that ρ_{ee} is upper bounded by 1/2, leading to a maximal value of $|\mathbf{F}_{\text{rad}}| \leq \hbar\mathbf{k}\Gamma/2$. This value is achieved with a large saturation parameter $s \equiv 2\Omega_\ell^2/\Gamma^2 \gg 1$. From the second line of the equation above, it can also be observed that the magnitude of light force decreases with detuning for a fixed decay rate and the Rabi frequency. This property is crucial for the application of optical molasses to cool atoms, as will be discussed in detail in Sec. 4.2.

The expression of Eq. (3.59) has a very clear physical meaning. In the presence of a laser light with wave-vector $\mathbf{k}$, an atom at ground state can absorb a photon from the laser and gain a momentum of $\hbar\mathbf{k}$. If the atom decays to the ground state via a stimulated emission process, momentum will be returned to the light field and no net effect is caused. However, if the atom undergoes a spontaneous decay, the emitted photon is in a random direction. Thus, the average over many spontaneous processes leads to a zero momentum change before and after the emission. As a result, the momentum $\hbar\mathbf{k}$ acquired from the light field is kept within the atom. Since the probability for a spontaneous emission to happen within a time interval dt is $\Gamma\rho_{ee}dt$, the momentum gained by the atom during the same time is

$$d\mathbf{p} = \hbar\mathbf{k}\Gamma\rho_{ee}dt, \tag{3.60}$$

which is consistent with the result of Eq. (3.59).

This part of light force $\mathbf{F}_{\text{rad}}$ is called the *radiation pressure force*. This force is not conservative, since the spontaneous decay is an irreversible process. Indeed, information of the absorbed photon from the single-mode laser light with fixed wave-vector and polarization is destroyed after the incoherent spontaneous emission process, where the emitted photon can have an arbitrary polarization and go in a random direction. Due to its dissipative nature, the radiation pressure force cannot be used to trap atoms. However, it plays an essential role in the cooling of atoms. In fact, through the incoherent processes of spontaneous emission, the entropy within an atomic ensemble can be taken away by the emitted photons, causing a decrease of temperature.

The first term in the light force of Eq. (3.57) corresponds to the energy shift discussed in Sec. 3.3.2, as one would expect from its relation to the gradient of the laser intensity. Indeed, if we consider the limiting case of a zero spontaneous decay rate $\Gamma \to 0$, the light force becomes

$$\mathbf{F}_{\text{dip}} = -\frac{2\hbar\delta\rho_{ee}}{\Omega_0}\nabla\Omega_0. \tag{3.61}$$

By employing the expressions for the eigenvalues Eq. (3.36) and eigenfunctions Eq. (3.37) of the atomic dressed states, it can be easily shown that the light force in this case is just the negative spatial gradient of the ground dressed state

$$\mathbf{F}_{\text{dip}} = -\nabla E_g'. \tag{3.62}$$

This part of the light force is called the *dipole force*, since it directly corresponds to the effect of an electric field exerted on a classical electric dipole. In the quantum mechanical picture, this force originates from the scattering processes of photons from one laser beam to another. Because the initial and final states of scattered photons are specified, these processes are completely reversible and the dipole force is conservative. In this sense, the dipole force can be used for trapping, but not for cooling of atoms. This can also be understood by noticing that the entropy of incident and scattered photons are both zero, such that no entropy can be carried away from the atom.

As a particular example, we consider a standing wave consisting of two counterpropagating laser beams. The Rabi frequency takes the form

$$\Omega = \Omega_\ell \cos(\mathbf{k} \cdot \mathbf{r}). \tag{3.63}$$

The dipole force Eq. (3.61) becomes

$$\mathbf{F}_{\text{dip}} = \frac{\hbar\delta\mathbf{k}\Omega_\ell^2 \sin(2\mathbf{k} \cdot \mathbf{r})}{4\delta^2 + \Gamma^2 + 2\Omega_\ell^2 \cos(\mathbf{k} \cdot \mathbf{r})}. \tag{3.64}$$

For the red-detuned laser with $\delta < 0$, the dipole force confines the atoms at positions with the highest laser intensity. For the blue-detuned laser with $\delta > 0$, the atoms are attracted to positions with the lowest laser intensity. As a result, the atoms would experience a periodic potential in a standing wave, which can be used to mimic a periodic lattice potential as in solid-state systems. The realization of the so-called *optical lattice* and its application in quantum emulation will be discussed in detail in Part III.

References

[1] C. Cohen-Tannoudji, J. Dupont-Roc and G. Crynberg. *Atom-Photon Inter-actions*. Wiley-VCH, Weinheim (2004).

[2] M. O. Scully and M. S. Zubairy. *Quantum Optics*. Cambridge University Press, Cambridge (1997).

[3] A. Einstein. Verhandlungen der Deutschen Physikalischen Gesellschaft **18**, 318 (1916).

[4] E. T. Jaynes and F. W. Cummings. *Proc. IEEE* **51**(1), 89 (1963).

[5] I. I. Rabi. *Phys. Rev.* **51**, 652 (1937).

[6] F. Bloch. *Phys. Rev.* **70**, 460 (1946).

4
Laser Cooling and Trapping

The experimental realization of cooling and trapping neutral atoms with the assistance of laser light is one of the most important achievements in cold atom physics. One of the original motivations for this pursuit is to achieve precision measurement. As we have learned in Chap. 2, the transition frequency between two atomic levels (usually two hyperfine states) can be used as frequency standards. However, a direct application of this proposal using atoms at room temperature would be compromised with displacement and broadening of the spectral lines induced by the Doppler shift of a wide spread of atomic velocities. Thus, it is of great importance to reduce the velocity distribution of atoms, or equivalently, to reduce the temperature of an atomic ensemble. Besides, the high atomic velocities also limit the observation time, and hence the spectral resolution, in an apparatus with a finite size. If one would be able to trap atoms within a confined space, the performance of measurement will be significantly enhanced.

The idea of cooling and trapping atoms via atom-light interaction was proposed by Askar'yan [1], Ashkin [2], Hänsch and Schawlow [3], and Wineland and Dehmelt [4]. The first demonstration of laser cooling was achieved in 1978 by Wineland *et al.* using Mg ions [5] and simultaneously by Neuhauser *et al.* using Ba ions [6]. After these pioneering works, the field of laser cooling and trapping underwent an enormous development, where at least several orders of magnitude improvements have been made in both temperature and atomic density. At present, the temperature can be reduced to as low as a few hundred pKs [7], which is the lowest temperature ever attained by mankind.

With the knowledge of atom-light interaction discussed in the previous chapter, we introduce some basic ideas about the cooling and trapping of neutral atoms in this chapter. Our goal is to give a brief overview of the experimental configurations that should be understood by anyone (even a theorist) who is interested in this field. For details regarding both experimental techniques and historical developments, readers are referred to the excellent book by Metcalf and van der Straten [8], the special issues edited by Phillips [9] and Chu and Wieman [10], and the review articles by Cohen-Tannoudji [11] and Adams and Riis [12].

4.1. Beam deceleration

Beam deceleration is a direct application of the radiation force $\mathbf{F}_{\mathrm{rad}}$ exerted by a laser beam on an atom, which corresponds to a momentum transfer between the photon and the atom. To see this clearly, we consider the simplest configuration where a laser beam incidents along the opposite direction to the atomic velocity $\mathbf{v}_0$, as depicted in Fig. 4.1. If the laser is nearly resonant with the transition frequency of atomic levels, the atom can absorb a photon and gain a recoil momentum of $\hbar\mathbf{k}$. The excited atom will return to the ground state either via a stimulated or a spontaneous decay process. For the stimulated case, the gained momentum $\hbar\mathbf{k}$ is returned to the laser, resulting in a zero change of atomic velocity. For the spontaneous case, however, the emitted photons are randomly oriented, such that the average effect over many emitting processes is zero. Thus, the recoil momentum is effectively absorbed by the atom, leading to a final velocity of $\mathbf{v}' = \mathbf{v}_0 + \hbar\mathbf{k}/m$ with m the atomic mass. Although the velocity change $|\hbar\mathbf{k}/m| \sim 1\,\mathrm{cm/s}$ of a single absorption-emission cycle is usually

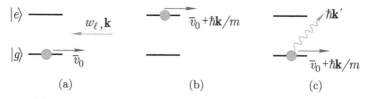

            (a)   (b)   (c)

Figure 4.1. (a) An atom at ground state with velocity $\mathbf{v}_0$ is subjected to a laser beam with frequency ω_ℓ and wave vector $\mathbf{k}$, which is antiparallel to the atomic velocity. The momentum associated with a laser photon is $\hbar\mathbf{k}$. (b) After absorbing a photon from the laser, the atom is slowed by $\hbar\mathbf{k}/m$. (c) After spontaneously emitting a photon in a random direction, the velocity of the atom is not changed on average, and is slower than that in (a).

very small compared to $|\mathbf{v_0}| \sim 10^2$ m/s for atoms at room temperature, the atomic velocity can be reduced significantly if the process can be repeated thousands of times.

In order to drive the atom through the absorption-emission cycle repeatedly and efficiently, one has to take two issues into account, the first being the optical pumping. In the picture we outlined above, the atomic levels are considered as a simplified two-level structure, such that atoms on the excited state $|e\rangle$ have nowhere to go but to return to the ground state $|g\rangle$. However, any realistic atom is not a two-level atom, but has a complicated spectrum consisting of many, sometimes degenerate, atomic levels. As a result, there is a chance for atoms on the excited state to emit a photon and relax to another state $|d\rangle$, which is far-off resonant from state $|e\rangle$. Atoms on the state $|d\rangle$ then become "dark" to the laser beam and can no longer go through the cooling process. One way to resolve this issue is to apply a second laser frequency, which drives the atoms in the dark state $|d\rangle$ back to the excited state $|e\rangle$, so that they can continue the absorption-emission cycle. This second laser is called the *repumper*, as illustrated in Fig. 4.2.

The second issue is the Doppler shift. In order for an atom to absorb photons from the laser beam efficiently, the laser frequency must be nearly resonant with the atomic transition frequency ω_{eg}. If an atom is moving with the initial velocity $\mathbf{v_0}$, the laser frequency will be Doppler-shifted such that the resonance condition is given by

$$\omega_{eg} \approx \omega_\ell - \mathbf{k} \cdot \mathbf{v_0}. \qquad (4.1)$$

After repeatedly going through the absorption-emission process, the atom is decelerated. This change of velocity alters the Doppler shift such that the atom is out of resonance with the light. In a typical case of a cold atom

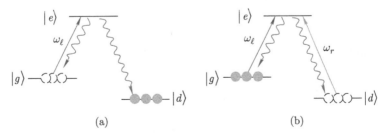

(a)   (b)

Figure 4.2.   (a) An atom can decay to an undesired dark state $|d\rangle$ such that the cycling transition between $|g\rangle$ and $|e\rangle$ is prevented. (b) A repumping laser of frequency ω_r is applied to bring atoms in the dark state back to the absorption-emission cycle.

experiment, a change in velocity of $\sim$5 m/s gives a large enough Doppler shift so that the absorption rate is significantly reduced. As a result, only atoms within a small window of velocity can be slowed down by the laser beam and they are only decelerated by a small amount. To solve this issue, the following three strategies can be used in practice.

• **Varying the laser frequency**

One way to keep the laser in resonance with the atomic transition frequency is to sweep the laser frequency ω_ℓ upward to compensate for the decreasing Doppler shift as the atoms are decelerated. This method is easy to implement since it does not require significant modification of the setup. The variation of frequency can be achieved by using semiconductor laser diodes, where a few mA sweep changes the laser frequency by several GHz, large enough to compensate for the effect of the Doppler shift [13]. One key feature of this so-called *chirp-cooling* technique is that the atoms are decelerated and arrive at the destination in pulses. This may be desired, of no importance or undesired, depending on the purpose of the experiments.

Another method is to use a broadband light, rather than a single-mode laser, to cover the frequency domain of $(\omega_{eg}, \omega_{eg} - \mathbf{k} \cdot \mathbf{v}_0)$. Then, any atom with a velocity below v_0 will be in resonance with some mode within this counterpropagating broadband light [14, 15]. One important disadvantage of this method is that only a small fraction of the light power is utilized to slow down the atoms, while the major component is effectively wasted.

• **Varying the atomic transition frequency**

As we have already learned in Sec. 2.4, the energy of atomic hyperfine states will be shifted in the presence of an external magnetic field. Thus, in a spatially varying magnetic field, the Zeeman shift can change the energy level separation in a desired way so that one can keep it in resonance with a laser of fixed frequency. In this *Zeeman cooling* method, a magnet (called the *Zeeman slower*) is carefully designed with more windings near the atom source and fewer windings near the end, so that the field strength is higher at the entrance and lower at the exit, to provide the appropriate Doppler shift along the path of moving atoms. A schematic illustration of the experimental configuration is shown in Fig. 4.3. Unlike the chirp-cooling method, atoms coming out of the Zeeman slower form a continuous beam. [16].

The atomic level separation can also be tuned using an inhomogeneous electric field via the d.c. Stark effect [17]. In this method, two long, oppositely charged plates are typically placed and separated by a tapered

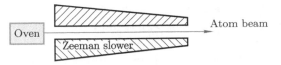

Figure 4.3.   A schematic representation of a Zeeman slower, which generates an inhomogeneous magnetic field to compensate for the change of Doppler shift during deceleration.

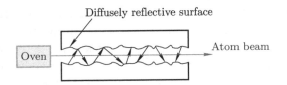

Figure 4.4.   A schematic illustration of diffusive light cooling apparatus.

gap ranging from one to a few centimeters, and charged to a few tens of kiloVolts. This technique features open access to the atomic beam along the lateral direction, which can be used for transverse cooling and/or collimation [18].

• **Varying the Doppler shift**

It is also possible to compensate the Doppler shift by noticing that there is an angular dependence in the term $\mathbf{k} \cdot \mathbf{v}$. If the decelerating laser beam is applied diffusively, rather than counterpropagating the atomic velocity, the Doppler shift experienced by the atom may vary in the range $(-kv, kv)$, depending on the relative angle θ between the atomic velocity and the light beam. The resonance condition thus can be fulfilled as the angle satisfies

$$\omega_\ell - \omega_{eg} = kv \cos \theta. \tag{4.2}$$

This method can be implemented via a tube with diffusely reflecting inner surface, as illustrated in Fig. 4.4. Because the diffusive light is isotropic, the atomic motion in any direction will undergo the same absorption-emission processes, hence it can be decelerated at the same time. This advantage helps prevent the atomic beam from being transversely expanded.

4.2.   Doppler cooling

In the previous section, we discuss the deceleration of atoms using a single laser beam. By implementing various techniques, the atomic motion along

one direction can be slowed by a laser beam due to the radiative force
exerted on atoms. This idea can be naturally extended to cases with more
than just one beam. For example, if two laser beams of the same frequency,
intensity and polarization are applied in directly opposite directions, an
atom moving along this line would experience radiative forces from both
beams. When the atom is at rest, the radiative force from each laser is
given by Eq. (3.59) with $\mathbf{F}_{\text{rad}} \propto \mathbf{k}$. The net force experienced by the atom
obviously vanishes because the wave-vectors of the two beams are oppo-
site. However, if the atom is moving along this line with some velocity
$\mathbf{v}$, the forces from different beams become uneven. Specifically, if both
lasers are red-detuned with frequency ω_ℓ less than the atomic transition
frequency, the frequency of the light beam opposing the atomic motion
is Doppler shifted toward the blue in the atomic rest frame, hence it is
brought close to resonance provided that the atomic velocity is not very
large. On the contrary, the frequency of the light pointing along the atomic
motion is Doppler shifted to the red and is further away from resonance.
As a consequence, the atom has a higher probability in absorbing photons from the opposing laser beam, leading to a net force which tends
to decelerate the atom. Because of the role of the Doppler effect in this
technique, this cooling mechanism is called *Doppler cooling*. This setup is
also referred to as *optical molasses* due to the similarity of the laser force
to viscous friction. A schematic illustration of Doppler cooling is shown in
Fig. 4.5.

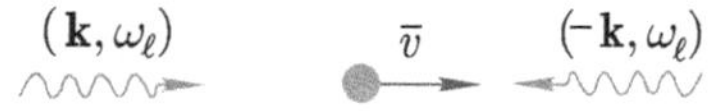

$$(\mathbf{k}, \omega_\ell) \qquad \bar{v} \qquad (-\mathbf{k}, \omega_\ell)$$

Figure 4.5.   A schematic illustration of a one-dimensional Doppler cooling. As the laser
frequency ω_ℓ is red-detuned, a right-moving atom is more likely to absorb photons from
the left-going laser, hence it can be decelerated.

To further demonstrate this mechanism, we consider the case where
the light intensity is low enough such that stimulated emission is not dramatic. This low-intensity limit allows us to ignore the processes where atoms
absorb photons from one beam and emit to the other. This type of transition
can lead to very large changes in atomic velocity. By considering spontaneous decay processes only and by employing the expression Eq. (3.59), the
net force exerted on an atom with velocity $\mathbf{v}_0$ can be written as

$$\mathbf{F}_{\text{Doppler}} = \mathbf{F}_+ + \mathbf{F}_-, \tag{4.3}$$

where

$$\mathbf{F}_{\pm} = \pm \frac{\hbar k \Gamma \Omega_\ell^2}{\Gamma^2 + 4(\delta \mp \mathbf{k} \cdot \mathbf{v}_0)^2 + 2\Omega_\ell^2}. \tag{4.4}$$

As the velocity $\mathbf{v}_0$ is small, the force takes the approximated form

$$\mathbf{F}_{\text{Doppler}} \approx \frac{16 \hbar k^2 \delta \Gamma \Omega_\ell^2}{(\Gamma^2 + 4\delta^2 + 2\Omega_\ell^2)^2} \mathbf{v}_0, \tag{4.5}$$

where higher order terms of $|\mathbf{v}_0|^2$ are neglected. Notice that for the red-detuned case of $\delta < 0$, the light force opposes the atomic motion and depends linearly on the velocity $\mathbf{v}_0$. This results in viscous damping of the atoms and this is how the term "optical molasses" originated. A typical velocity dependence of $|\mathbf{F}_{\text{Doppler}}|$ is shown in Fig. 4.6.

The same idea can be easily generalized to three dimensions with atoms loaded at the intersection of three pairs of counter propagating laser beams, and the motion of atoms in all directions can be severely restricted [19]. We emphasize that an optical molasses is not a trap. The velocity-dependent light force is only a viscous damping force, but not a restoring force which should depend on spatial positions.

The lowest temperature attainable via Doppler cooling can be estimated by equating the cooling rate to the heating rate. The cooling rate is $\gamma_C = \mathbf{F}_{\text{Doppler}} \cdot \mathbf{v} \propto v^2$. To estimate the heating rate, we notice that as the atom spontaneously emits a photon in a random direction, the average velocity change of the atom is zero, but the root mean square (RMS) is finite. One can imagine that this finite change of RMS causes a heating effect to the atom. The heating rate is thus proportional to the photon recoil energy and the total scattering rate $\gamma_H \propto E_R \Gamma \rho_{ee} \propto \hbar^2 k^2 \Gamma \rho_{ee}/2m$. By substituting the expressions for $\mathbf{F}_{\text{Doppler}}$ and ρ_{ee} into the equilibrium

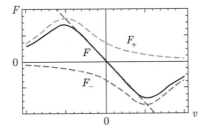

Figure 4.6. Velocity dependence of the optical damping forces for one-dimensional optical molasses.

condition $\gamma_C = \gamma_H$, we obtain the temperature associated with the kinetic energy of the steady-state

$$k_B T_D \propto \frac{\Gamma^2 + 4\delta^2 + 2\Omega_\ell^2}{8\delta}, \tag{4.6}$$

where k_B is the Boltzmann's constant and T_D is called the *Doppler temperature* or the *Doppler cooling limit*. In the low-intensity limit with $\Omega_\ell \ll (\Gamma, |\delta|)$, the lowest value of Eq. (4.6) is obtained as $4\delta^2 = \Gamma^2$, leading to a Doppler temperature $T_D \propto \hbar\Gamma/2k_B$. For atomic transitions that are usually used in practical cold atom experiments, the Doppler temperature T_D is below 1 mK.

4.3.   Evaporative cooling

In the previous section, it is shown that a lower limit to the Doppler laser cooling exists, at which point the cooling effect is completely compensated by the heating effect induced by the randomness of spontaneous emission. Considering the fact that the spontaneous emission is irreversible, this heating mechanism is inevitable in any laser cooling schemes. In order to go below the Doppler limit, the heating rate is to be reduced by carefully tailoring a spatially dependent optical pumping process, such that a moving atom seems to be always climbing potential energy hills and losing kinetic energy [20, 21]. This scheme obviously brings to mind a Greek myth and this process is thus called *Sisyphus laser cooling*. The spatially dependent optical pumping processes can be achieved in several different ways, including two counter propagating laser beams having orthogonal polarizations [20, 21] (see Fig. 4.7), a standing wave of constant polarization and an additional DC magnetic field [22, 23], and an optical molasses at high intensity [24].

Even with the aforementioned sub-Doppler cooling techniques, there is still a limitation of the minimum steady-state value of the average kinetic energy. This temperature limit is of the order of the photon recoil energy $T_R = E_R/k_B$ and is called the *recoil limit*. In fact, one may expect that the last spontaneous emission in a cooling process would leave atoms with a residual momentum of the order of $\hbar k$. Since this recoil momentum is in a random direction, a distribution of atomic velocity with a finite uncertainty is expected. This rough estimation is confirmed by a more delicate quantitative analysis [25], which shows that the narrowest velocity distribution attainable by the Sisyphus cooling corresponds to a

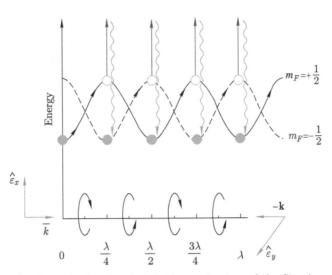

Figure 4.7. A schematic drawing depicts the mechanism of the Sisyphus cooling. In this scheme, two linearly polarized counterpropagating laser beams with orthogonal polarizations form a standing wave. When an atom is moving along the standing wave, it oscillates between the $m_F = 1/2$ and $m_F = -1/2$ states. As a result, the atom always climbs potential energy hills and is slowed during this process.

temperature $\sim 10 T_R \sim \mu$K. Although the recoil limit can be surpassed by a few clever schemes, it is generally regarded as the lower limit for all optical cooling methods.

Another route to further cool the atomic gases, especially below the recoil limit, is by implementing a technique called *evaporative cooling*. This method is based on the preferential removal of those atoms possessing higher energy than the average energy, followed by a rethermalization of the remaining atoms by elastic collisions. This idea is very simple and can be encountered in our everyday life. Indeed, the major mechanism for a cup of hot coffee to be cooled down to room temperature is by evaporation. During that process, the most energetic water molecules evaporate from the cup, while the remaining molecules collide with each other to regain thermal equilibrium and the cup of coffee is cooled to a lower temperature. Furthermore, our daily experience tells us that only a small fraction of molecules are required to be evaporated in order to cool the cup of coffee by a considerable amount. As we will see, this property guarantees the success of evaporative cooling in obtaining a low temperature and a high phase space density simultaneously.

To gain some insight into the evaporative cooling process without resorting to detailed calculations, we consider a simple example of atoms confined in a three-dimensional isotropic harmonic trap in the following equation

$$U(\mathbf{r}) = U_0 \left(\frac{r}{a}\right)^2, \tag{4.7}$$

where U_0 represents the trap depth and a is the characteristic length. The evaporation can be initiated by lowering the trap depth, hence allowing the atoms with energies higher than the new depth to escape. A schematic illustration of this procedure is shown in Fig. 4.8. Although in practical cases, the trap depth is usually lowered continuously according to an optimized cooling trajectory. Here, we analyze a simplified model where the trap is lowered in one single step [26–28].

Consider a total number of N atoms initially loaded in an infinitely deep trap with a temperature T. Assuming the gas is described by classical statistics, the total number of atoms in the trap satisfies the number equation

$$N = \int_0^\infty D(E) e^{-(E-\mu)/k_B T} dE, \tag{4.8}$$

where the exponential factor stems from the Maxwell–Boltzmann distribution, μ is the chemical potential and $D(E)$ is the density of states

$$D(E) = \frac{2\pi (2m)^{3/2}}{(2\pi\hbar)^3} \int \sqrt{E - U(\mathbf{r})} d^3\mathbf{r}. \tag{4.9}$$

After lowering the trap depth to a finite value $U' = \eta k_B T$, all atoms with higher energies escape from the trap. The number of atoms remaining in

Figure 4.8.   Illustration of the evaporative cooling process. Once the trap depth is lowered, atoms with energy above the trap depth can escape from the trap and the remaining atoms rethermalize to reach a lower temperature.

the trap becomes

$$N' = \int_0^{\eta k_B T} D(E) e^{-(E-\mu)/k_B T} dE. \tag{4.10}$$

By solving the initial number Eq. (4.8) for the chemical potential μ and substituting its expression into Eq. (4.10), we obtain

$$\frac{N'}{N} = \int_0^{\eta} \Delta(\epsilon) e^{-\epsilon} d\epsilon, \tag{4.11}$$

where the reduced energy is defined as $\epsilon \equiv E/k_B T$ and the reduced density of the states is given by $\Delta(\epsilon) = \epsilon^2/2$. The integral in Eq. (4.11) can be evaluated analytically to give

$$\frac{N'}{N} = \frac{2 - (\eta^2 + 2\eta + 2)e^{-\eta}}{2}. \tag{4.12}$$

Notice that the fraction of atoms remaining in the trap is solely determined by the final trap depth η.

The average reduced energy $\bar{\epsilon}$ of the atoms before the cooling process is given by

$$\bar{\epsilon} = \frac{\int_0^{\infty} \epsilon \Delta(\epsilon) e^{-\epsilon} d\epsilon}{\int_0^{\infty} \Delta(\epsilon) e^{-\epsilon} d\epsilon} = 3. \tag{4.13}$$

The average reduced energy $\bar{\epsilon}'$ after the cooling process can be obtained using the same expression by changing the integral upper boundary from ∞ to η, leading to

$$\bar{\epsilon}' = 3 + \frac{\eta^3}{\eta^2 + 2\eta + 2 - 2e^{\eta}}. \tag{4.14}$$

Since the average energy is proportional to the temperature, the resulting temperature after evaporation is given by[1]

$$\frac{T'}{T} = \frac{\bar{\epsilon}'}{\bar{\epsilon}} = 1 + \frac{\eta^3}{3(\eta^2 + 2\eta + 2 - 2e^{\eta})}. \tag{4.15}$$

---

[1] According to thermodynamics, temperature is defined as a Lagrange multiplier in a canonical ensemble that is in thermal equilibrium with a heat bath. In cold atom physics, however, the atoms are carefully kept isolated from the environment to avoid heat transport. Thus, it is not applicable in principle to define a *temperature* in such an isolated system. Instead, the meaningful physical quantity here is the internal energy. In the thermodynamic limit, it can be proven that the internal energy for a micro canonical ensemble is proportional to the temperature for a canonical ensemble.

Following the same procedure, we can also obtain the change of density, and most importantly, of phase-space density after the evaporation process. In fact, a simple dimension analysis shows that the volume occupied by trapped atoms scales as $V \propto T^{3/2}$. As a result, the real-space number density $n \equiv N/V \propto NT^{-3/2}$, while the phase-space density $\rho \equiv n\lambda_{\mathrm{dB}}^3 \propto NT^{-3}$, with λ_{dB} the de Broglie wavelength. From those, we obtain the following expressions for n'/n and ρ'/ρ,

$$\frac{n'}{n} = \left(\frac{N'}{N}\right)\left(\frac{T}{T'}\right)^{3/2},$$

$$\frac{\rho'}{\rho} = \left(\frac{N'}{N}\right)\left(\frac{T}{T'}\right)^{3}. \tag{4.16}$$

In Fig. 4.9, we show the results for the fraction of atoms remaining in the trap, the temperature ratio, the real-space density ratio and the phase-space density ratio as functions of the final trap depth η. The most intriguing feature is that the phase-space density increases monotonically

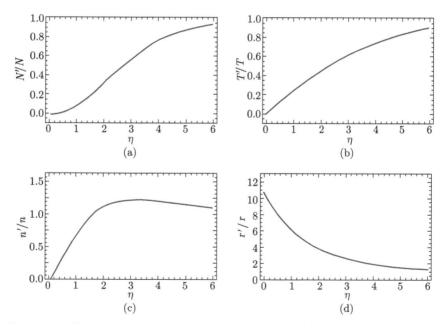

Figure 4.9. Results for the one-step evaporative cooling of (a) fraction of atoms remaining in the trap, (b) temperature ratio, (c) real-space density ratio and (d) phase-space density ratio as functions of the final trap depth η.

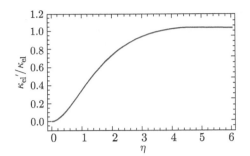

Figure 4.10.   Results of the elastic collision rate ratio $\kappa'_{\mathrm{el}}/\kappa_{\mathrm{el}}$ as a function of η.

with decreasing η and exceeds the critical value of 2.612 for the onset of the Bose–Einstein condensation (BEC) at a fairly large value of $\eta \approx 2.9$.

The success of evaporative cooling depends crucially on the effect of rethermalization of the remaining atoms. Elastic collisions are responsible for this to happen. Indeed, experimental results show that ~ 2.7 elastic collisions are necessary to rethermalize the gas [29]. In the low energy limit, the cross section for elastic scattering is dominated by the s-wave contribution and is given by $\sigma = 8\pi a_s^2$, where a_s is the s-wave scattering length. Thus, the elastic collision rate $\kappa_{\mathrm{el}} \equiv n\sigma v \propto nT^{1/2}$. In Fig. 4.10, we plot the result for $\kappa'_{\mathrm{el}}/\kappa_{\mathrm{el}}$ as a function of η. Notice that the elastic collision rate nearly remains a constant for $\eta \gtrsim 3$ and drops almost linearly for $\eta \lesssim 3$. The high elastic collision rate during evaporative cooling is indispensable for the technique to be applicable.

The first attempt to apply evaporative cooling to cool atoms was carried out in 1988 [30]. In the following two decades, this method proved to be very powerful due to its simplicity for implementation, wide applicability to different atom species, and most importantly, high efficiency for increasing phase-space density. Indeed, in all the earliest experiments that achieved BEC, evaporative cooling was the key ingredient to increase the phase-space density by about six orders of magnitude. In Table 4.1, the results of evaporative cooling from the first three groups that have obtained BEC are summarized.

4.4.   Magnetic trapping

In the previous sections, we have introduced several methods to cool an ensemble of neutral atoms. After the atoms have been cooled, we want

Table 4.1.   Results of evaporative cooling from the first three groups that
have obtained BEC. The first and second lines in each experiment represent the initial and final conditions of the evaporation. The phase-space
density needs to exceed the critical value of 2.612 to achieve BEC.

Group	Atom	N (10^6)	n $(10^{12}\,\mathrm{cm}^{-3})$	T $(\mu\mathrm{K})$	ρ (10^{-6})
Rice [31]	^{7}Li	200	0.07	200	7
		0.1	1.4	0.4	
MIT [32]	^{23}Na	1000	0.1	200	2
		0.7	150	2	
JILA[a] [33]	^{87}Rb	4	0.04	90	0.3
		0.02	3	0.17	

[a]The Joint Institute for Laboratory Astrophysics.

them to be stored at a fixed location for a fairly long period of time, so that
subsequent manipulation or measurement can be performed. In the case of
charged particles, ion trapping, laser cooling of trapped ions and trapped
ion spectroscopy were known for many years [36]. For neutral atoms, the
trapping schemes heavily rely on the interaction between an inhomogeneous
electromagnetic field and an atomic multipole moment. Since atoms in the
s-wave ground state do not acquire an electric dipole moment due to their
inversion symmetry, the first idea one would come up with is to use magnetic
traps, which couple to the atomic magnetic dipole moments.

Consider a neutral atom loaded in an inhomogeneous magnetic field
$\mathbf{B}(\mathbf{r})$. When the atom is at rest, as we have learned in Sec. 2.4, the hyperfine
states will be split due to the Zeeman effect and the resulting energies
become field dependent (see for example, Fig. 2.3)

$$E = E_0 - \mu B(\mathbf{r}) + \mathcal{O}(B^2). \qquad (4.17)$$

Here, the coefficient μ of the linear Zeeman term represents the magnetic
dipole moment. The hyperfine states with $\mu > 0$ are driven to positions
of higher magnetic field, hence are called *high-field seekers*. The hyperfine
states with $\mu < 0$ are driven to positions of lower magnetic field and are
referred as *low-field seekers*.

When the atom is moving, the magnetic field experienced by the atom
becomes time dependent and in principle can induce transitions between
different states. A general treatment of this process then requires a full
solution of the time-dependent Schrödinger equation. However, if the atom
travels slowly enough and the gap between the ground and excited atomic
states is large enough, the transition amplitude is negligible and the internal

state of the atom is assumed to change instantaneously with the magnetic field. This limit is usually called *adiabatic approximation*. Since the time scale associated with the atomic orbital motions is the cyclotron frequency ω_T, while the time scale of the transition between different hyperfine states is the Lamor precession frequency ω_Z, the validity of adiabatic approximation is governed by the following condition

$$\omega_T = \frac{1}{B}\left|\frac{dB}{dt}\right| \ll \omega_Z = \frac{\mu B}{\hbar}. \tag{4.18}$$

When this condition is satisfied, the inhomogeneous magnetic field then provides an effective potential energy

$$U_{\text{eff}}(\mathbf{r}) = -\mu B(\mathbf{r}). \tag{4.19}$$

According to this expression, if we put a low-field seeker in an inhomogeneous magnetic field, it will be trapped by a local field minimum.[2]

One way to create a local magnetic field minimum is by arranging two identical coils carrying opposite currents to set up a so-called *quadrupole trap*, as shown in Fig. 4.11. This trap clearly has a single minimum at the trap center where the magnetic field reaches zero. Around the trap center, the magnitude of the field is given by

$$B(\mathbf{r}) = B_0\sqrt{x^2 + y^2 + 4z^2}, \tag{4.20}$$

where the z-axis is denoted to be along the axial direction of the coils and B_0 is a constant determined by the current and the geometric configuration of the trap [37]. Therefore, the magnitude of the field varies linearly with distance from the minimum, with a slope that depends on direction.

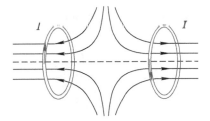

Figure 4.11. A schematic illustration of a quadrupole trap. The magnetic field reaches its minimum of zero at the trap center.

---

[2]A local maximum of magnetic field is prohibited by Earnshaw's theorem.

The most attractive aspect of a quadrupole trap is its experimental simplicity. It is easy to construct and has optical access in all three directions. However, a quadrupole trap also suffers from the important disadvantage of the zero magnitude of magnetic field at the trap center. From Eq. (4.18), we can see clearly that a large enough magnetic field is a prerequisite to fulfill the adiabatic condition that guarantees the validity of the discussion above. This is certainly not the case at the center of a quadrupole trap. In fact, if an atom initially in a low-field seeker state reaches the trap center, the gap between different hyperfine states vanishes and the energy of some low-field and high-field seekers become degenerate. Then, there is a significant probability that the atom can be transferred to a high-field seeker state and pushed away from the trap. This mechanism results in appreciable atom loss in the vicinity of the zero-field point, i.e., a "hole" near the center of a quadrupole trap.

There are several ways to circumvent this disadvantage and plug the hole. One way is to apply an additional oscillating magnetic field within the radial x-y plane

$$B_{\text{osc}}(t) = B_b \cos(\omega t)\hat{x} + B_b \sin(\omega t)\hat{y}, \qquad (4.21)$$

where ω is the oscillating frequency. The time-averaged effect of this oscillating field is to introduce a bias potential to the quadrupole trap

$$\langle B(\mathbf{r}) \rangle_t = B_b + \frac{B_0^2}{4B_b} \left(x^2 + y^2 + 8z^2 \right). \qquad (4.22)$$

This modified quadrupole trap is called the *time-averaged orbiting potential* (TOP) trap. Notice that the TOP trap does not have any zero-field point. Besides, the presence of the oscillating field also modifies the linear behavior of the quadrupole trap to a quadratic one, hence generating an anisotropic harmonic potential. The trapping frequencies in the radial and axial directions are

$$\omega_x^2 = \omega_y^2 = -\frac{\mu B_0^2}{2mB_b},$$
$$\omega_z^2 = -\frac{4\mu B_0^2}{mB_b}. \qquad (4.23)$$

Since the axial confinement is stronger than the radial confinement with $\omega_z > \omega_{x,y}$, the TOP trap is of a pancake shape.

Another setup that creates a magnetic trap without any holes is the *Ioffe–Pritchard* trap, which is comprised of four straight currents and a pair

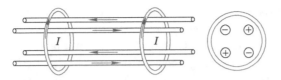

Figure 4.12.   A schematic illustration of an Ioffe–Pritchard trap. The four straight currents provide a magnetic gradient within the transversal plane, while the end coils close the trap along the axial direction.

of Helmholtz coils as shown in Fig. 4.12. A simple calculation shows that the four straight currents (so-called *Ioffe bars*) create a harmonic magnetic trap within the radial plane. The presence of the two Helmholtz coils will close the trap along the axial direction, provided that the distance between them is larger than the coil radius. The total magnetic field around the trap center takes the form

$$B(\mathbf{r}) = A_1 + \left(\frac{C^2}{2A_1} - \frac{3A_3}{2} \right)(x^2 + y^2) + 3A_3 z^2, \qquad (4.24)$$

where the coefficients A_1 and A_3 are determined by the Helmholtz coils, and the coefficient C is controlled by the Ioffe bars. Thus, if the currents in the Ioffe bars are strong enough to give $C^2 > 3A_1 A_3$, this setup generates a harmonic magnetic trap. The geometry of the Ioffe–Pritchard trap can be adjusted between a pancake shape and cigar shape by varying the parameters.

To conclude this section, we emphasize that for neutral atoms in the ground state, the magnetic dipole moments are quite weak and are on the order of a few Bohr magnetons. As a consequence, magnetic traps of all types are very shallow in practical experiments. To realize a successful trapping, the atoms must be cooled substantially to reach temperatures as low as a few mK. The small depth of magnetic traps also dictates a stringent requirement for inelastic collision, through which process atoms can be scattered out of the trap. This consideration sets up an upper limit for the number density of atoms. Besides, one also needs a very good vacuum chamber to protect the trapped atoms from destructive collisions with background gas.

4.5.   Optical trapping

In Sec. 3.5, we have shown that an atom will experience a dipole force, which is proportional to the gradient of the light shift induced by the atom-laser interaction. If the laser is red-detuned from the atomic transition frequency, the dipole force is in the same direction of the gradient of the light shift, hence it pushes the atoms to the point of maximal laser intensity and traps them nearby. This trapping effect is due to the interaction between the laser field and the atom electric dipole moment. The resulting trap is then usually called an *optical dipole trap.*

The simplest example of an optical dipole trap consists of a single, strongly focused Gaussian laser beam [34, 35]. The distribution of the laser intensity within the transversal plane around the focal point takes the following form

$$I(\rho) = I_0 e^{-\rho^2/w^2}, \tag{4.25}$$

where ρ is the radial coordinate within the plane perpendicular to the laser wave vector, w is the beam waist width and I_0 is the maximal intensity. An illustration of such a laser beam is shown in Fig. 4.13.

Within such a Gaussian beam, the dipole force experienced by an atom is given by Eq. (3.57), leading to

$$\mathbf{F}_{\text{dip}} = -\frac{s}{1+s}\frac{\hbar\delta\nabla\Omega_\ell}{\Omega_\ell} = -\frac{2\hbar\delta\Omega_\ell\nabla\Omega_\ell}{\Gamma^2 + 4\delta^2 + 2\Omega_\ell^2}. \tag{4.26}$$

In the large detuned limit with $|\delta| \gg \Omega, \Gamma$, the dipole force takes the form

$$\mathbf{F}_{\text{dip}} = -\frac{\hbar}{4\delta}\nabla\left(\Omega_\ell^2\right) \propto -\frac{\hbar}{4\delta}\nabla I(\rho)$$

$$\propto \frac{\hbar}{4\delta}\frac{\rho}{w^2}I_0 e^{-\rho^2/w^2}. \tag{4.27}$$

In the longitudinal direction, a more careful analysis shows there is also an attractive force depending on the details of the focusing. Thus, this trap produces an attractive force on atoms in all directions.

In addition to the dipole force, the atom-light interaction also results in a radiation pressure force as given in Eq. (3.59). Although this force is dissipative, it can also be employed to confine atoms in combination of an

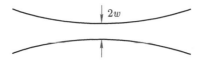

Figure 4.13.   A single focused laser beam produces a simple example of an optical dipole trap.

inhomogeneous magnetic field. This hybrid system with both optical and magnetic setups is called the *magneto-optical trap* (MOT). The mechanism of a MOT relies on selective optical pumping that is tuned by a magnetic field gradient. As an example, we consider the one-dimensional case with a linearly inhomogeneous magnetic field varying along the x-direction

$$B(x) = Ax. \tag{4.28}$$

Because of the Zeeman effect, the atomic transition frequencies between different hyperfine states also become spatial-dependent, as illustrated in Fig. 4.14. Here, we consider the simple scheme of $|F = 0, m_F = 0\rangle \rightarrow |F = 1, m_F = 0, \pm 1\rangle$ transitions. Two counterpropagating laser beams are applied along the x-direction with opposite circular polarizations. The beam from the left has σ_+ polarization, while the one from the right is σ_- polarized. The laser frequency, shown as dashed line in Fig. 4.14, is chosen to be red-detuned from the atomic transitions at a zero magnetic field.

In this configuration, if an atom is at the trap center, the $m_F = \pm 1$ states are degenerate and are equally detuned from the laser frequency. Thus, it is of equal probability for an atom to be scattered by the right-going σ_+ beam and the left-going σ_- beam. The total radiation pressure force exerted on the atom is zero. If an atom travels to the right of the trap center, e.g., at position x_1, the $|F = 1, m_F = -1\rangle$ state is shifted downwards and becomes nearly resonant with the laser. In this situation,

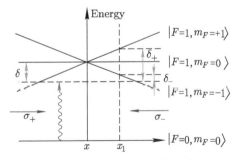

Figure 4.14.   Atomic transitions in a one-dimensional MOT. For an atom located at position x_1, the $m_F = -1$ state is shifted downwards to be nearly resonant with the laser frequency (dashed line), due to the Zeeman effect induced by a linear magnetic field gradient. Meanwhile, the $m_F = +1$ state is shifted upwards and becomes far detuned with $\delta_+ > \delta > \delta_-$. Thus, the atom is more likely to scatter a σ_- photon from the left-going laser and obtains a recoil momentum to the left. The net effect of this selective optical pumping is a restoring force toward the trap center.

the $|F = 0, m_F = 0\rangle$ ground state is more likely to absorb a σ_- photon from the left-going beam and obtains a recoil momentum to the left. The total radiation pressure force is then pointing toward the trap center. On the other side of the trap center, the $|F = 1, m_F = +1\rangle$ is nearly resonant to the laser frequency and the atom will scatter more photons from the σ_+ beam, resulting in a net force also toward the trap center.

In comparison to magnetic traps and optical dipole traps, the MOT has many advantages. First, since the trapping force is from the radiation pressure, a MOT is usually much deeper than a pure magnetic trap that relies on magnetic dipole coupling. Second, a MOT does not require precise balancing of the counterpropagating laser beams nor a high magnetic field gradient. It is then easy to construct a MOT with simple, air-cooled magnetic coils. Furthermore, the atoms are cooled simultaneously when they are trapped by the laser beams, so that a MOT can be operated, even starting from room temperature.

References

[1] G. A. Askar'yan. *Sov. Phys. JETP* **15**, 1088 (1962).
[2] A. Ashkin. *Phys. Rev. Lett.* **25**, 1321 (1970).
[3] T. Hänsch and A. Schawlow. *Opt. Commun.* **13**, 68 (1975).
[4] D. Wineland and H. Dehmelt. *Bull. Am. Phys. Soc.* **20**, 637 (1975).
[5] D. Wineland, R. Drullinger and F. Walls. *Phys. Rev. Lett.* **40**, 1639 (1978).
[6] W. Neuhauser, M. Hohenstatt, P. Toschek and H. Dehmelt. *Phys. Rev. Lett.* **41**, 233.
[7] P. Medley, D. M. Weld, H. Miyake, D. E. Pritchard and W. Ketterle. *Phys. Rev. Lett.* **106**, 195301 (2011).
[8] H. J. Metcalf and P. van der Straten. *Laser Cooling and Trapping*. Springer-Verlag, New York (1999).
[9] W. D. Phillips. *Prog. Quant. Elect.* **8**, 115 (1984).
[10] S. Chu and C. Wieman. *J. Opt. Soc. Am. B* **6**, 1961 (1989).
[11] C. Cohen-Tannoudji. *Phys. Rep.* **219**, 153 (1992).
[12] C. S. Adams and E. Riis. *Prog. Quant. Elect.* **21**, 1 (1997).
[13] R. Watts and C. Wieman. *Opt. Lett.* **11**, 291 (1986).
[14] M. Zhu, C. W. Oates and J. L. Hall. *Phys. Rev. Lett.* **67**, 46 (1991).
[15] I. C. M. Littler, H. M. Keller, U. Gaubatz and K. Bergmann. *Z. Phys. D* **18**, 307 (1991).
[16] W. Phillips and H. Metcalf. *Phys. Rev. Lett.* **48**, 596 (1982).
[17] R. Gaggl, L. Windholz, C. Umfer and C. Neureiter. *Phys. Rev. A* **49**, 1119 (1994).
[18] J. R. Yeh, B. Hoeling and R. J. Knize. *Phys. Rev. A* **52**, 1388 (1995).

[19] S. Chu, L. Hollberg, J. Bjorkholm, A. Cable and A. Ashkin. *Phys. Rev. Lett.* **55**, 48 (1985).

[20] J. Dalibard and C. Cohen-Tannoudji. *J. Opt. Soc. Am. B* **6**, 2023 (1989).

[21] P. J. Ungar, D. S. Weiss, S. Chu and E. Riis. *J. Opt. Soc. Am. B* **6**, 2058 (1989).

[22] B. Sheehy, S. Q. Shang, P. van der Straten, S. Hatamian and H. Metcalf. *Phys. Rev. Lett.* **64**, 858 (1990).

[23] D. S. Weiss, E. Riis, S. Chu, P. J. Ungar and Y. Shevy. *J. Opt. Soc. Am. B* **6**, 2072 (1989).

[24] J. Darlibard and C. Cohen-Tannoudji. *J. Opt. Soc. Am. B* **2**, 1707 (1985).

[25] Y. Castin, J. Dalibard and C. Cohen-Tannoudji. *Light Induced Kinetic Effects on Atoms, Ions, and Molecules.* Eds. L. Moi, S. Gozzini, C. Gabbanini, E. Arimondo and F. Strumia. Pisa, ETS Editrice (1991).

[26] K. B. Davis, M. O. Mewes and W. Ketterle. *App. Phys. B* **60**, 155 (1995).

[27] W. Ketterle and N. J. van Druten. *Adv. Atom. Mol. Opt. Phys.* **37**, 181 (1996).

[28] V. Bagnato, D. E. Pritchard and D. Kleppner. *Phys. Rev. A* **35**, 4354 (1987).

[29] C. Monroe, E. Cornell, C. Sackett, C. Myatt and C. Wieman. *Phys. Rev. Lett.* **70**, 414 (1993).

[30] N. Masuhara, J. M. Doyle, J. C. Sandberg, D. Kleppner, T. J. Greytak, H. F. Hess and G. P. Kochanski. *Phys. Rev. Lett.* **61**, 935 (1988).

[31] C. C. Bradley, C. A. Sackett, J. J. Tollett and R. G. Hulet. *Phys. Rev. Lett.* **78**, 985 (1997).

[32] K. B. Davis, M.-O. Mewes, M. R. Andrews, N. J. van Druten, D. S. Durfee, D. M. Kurn and W. Ketterle. *Phys. Rev. Lett.* **75**, 3969 (1995).

[33] M. H. Anderson, J. R. Ensher, M. R. Matthews, C. E. Wieman and E. A. Cornell. *Science*, **269**, (1995).

[34] A. Ashkin. *Phys. Rev. Lett.* **24**, 156 (1970).

[35] S. Chu, J. Bjorkholm, A. Ashkin and A. Cable. *Phys. Rev. Lett.* **57**, 314 (1986).

[36] D. Wineland, W. Itano, J. Bergquist and J. Bollinger. Technical Report 1086, N.I.S.T. (1985).

[37] T. Bergeman, G. Erez and H. Metcalf. *Phys. Rev. A* **35**, 1535 (1987).

5
Interaction Between Atoms

One of the most exciting areas of research on ultracold atomic gases is to emulate theoretical models and demonstrate general physics in multi-disciplinary fields. Toward this goal, efforts are mainly devoted along two directions. The first one is to manipulate the single-particle properties of an atom. This includes adjusting the single-particle dispersion via an optical lattice and introducing a synthetic gauge field by Raman transitions. The other direction is to investigate the interaction effect. Along this line, the realization of the Bose–Hubbard model, the study of the crossover between the Bardeen–Cooper–Shrieffer (BCS) and the BEC limits in Fermi gases, and the demonstration of Efimov physics shed light on our understanding of the corresponding many-body and few-body physics.

To study the interaction effect in ultracold atomic gases, we should bear in mind that the system is usually in the dilute limit where the inter-atomic distance d is much larger than the characteristic length scale of atomic interaction range R_e. Indeed, for a typical experimental situation with number density $n \sim 10^{14}$ cm^{-3}, the mutual distance between two atoms is $d \sim 10^2$ nm. As a comparison, the interaction range R_e is on the scale of an atomic radius, which is about several nanometers. The diluteness condition $d \gg R_e$ suggests that it is unlikely for three or more particles to come within close range of R_e, and hence the interaction effect is dominated by two-body processes. Besides, this condition also ensures that the short-range potential details are irrelevant for the low-energy behavior of the two-body processes.

In this chapter, we focus on two-body problems of neutral atoms. In Sec. 5.1, we give a brief introduction regarding the actual interaction potential between atoms. Since the particles do not carry net charges, the long-distance behavior of the interaction is dominated by the induced dipole interaction and takes the form of van der Waals potential $\sim r^{-6}$ in the large separation regime $r \to \infty$. This long-distance tail governs the low-energy and long-wavelength physics of two-body processes, hence it is of great interest in the discussion of atom gases at low temperatures. In Sec. 5.2, we discuss the elastic scattering between two atoms in free space. In the scattering theory, the internal states of the two colliding particles in the initial and final states are characterized by a set of quantum numbers. A possible choice of these quantum numbers is usually referred to as a *channel*. In this chapter, we discuss only the case where the internal states of the two colliding atoms do not change, i.e., the scattering takes place within a single channel. Multi-channel processes and the important consequence of the Feshbach resonance will be studied in Chap. 6. We emphasize that the full scattering process, even in the low-energy limit, is indeed a very complicated problem. Here, we only consider a very simple case, in order to introduce some basic but important concepts which will be useful in the rest of the book. A more complete discussion of the full story can be found in other places, see for example, the excellent book by J. R. Taylor [1].

5.1.   Interaction potential between alkali-metal atoms

In the most general case, the interaction between two atoms consists of two parts: a repulsive core resulting from the Pauli exclusion principle of the electrons, which prevents the two atoms from being located at the same position simultaneously, and a long-distance tail originating from electric or magnetic coupling between permanent or induced multipoles. Considering the fact that the nuclear multipole moments are much weaker, the long-distance component of the interatomic interaction has four major contributions from the electronic states:

- Electrostatic interactions between permanent charges, electric dipoles and higher-order permanent multipoles;
- Magnetic interactions between permanent magnetic dipoles and higher-order multipoles;

- Interaction between the permanent multipole of one atom and the induced multipole of another atom;
- An attractive interaction arising from the coupling of induced multipoles.

In this section, we consider the most relevant case of two neutral alkali-metal atoms in the electronic ground s-wave state, where the electric multipole moments are absent. Besides, since the permanent magnetic dipole moment of alkali-metal atoms is small ($\sim \mu_B$), the leading term of the interatomic interaction is rooted in induced electric dipole moments.

To derive the interaction potential between two atoms at a large distance, we note that the interacting Hamiltonian of two electric dipole moments $\mathbf{d}_1$ and $\mathbf{d}_2$ takes the form

$$
\begin{aligned}
H_{dd} &= \frac{1}{4\pi\epsilon_0 r^3} \left[\mathbf{d}_1 \cdot \mathbf{d}_2 - 3(\mathbf{d}_1 \cdot \hat{\mathbf{r}})(\mathbf{d}_2 \cdot \hat{\mathbf{r}}) \right] \\
&= \frac{1}{4\pi\epsilon_0 r^3} (d_{1x}d_{2x} + d_{1y}d_{2y} - 2d_{1z}d_{2z}),
\end{aligned} \tag{5.1}
$$

where $\mathbf{r}$ is the relative coordinate between the two dipoles. Since the ground state wavefunction associated with each atom is s-wave in nature, the first-order perturbation matrix element of H_{dd} vanishes. The second-order perturbation

$$
E^{(2)}(r) = \frac{1}{(4\pi\epsilon_0)^2 r^6} \sum_{(n_1,n_2)\neq(0,0)} \frac{|\langle n_1, n_2 |(d_{1x}d_{2x} + d_{1y}d_{2y} - 2d_{1z}d_{2z})|0,0\rangle|^2}{2E_0 - E_{n_1} - E_{n_2}}
$$

$$\tag{5.2}$$

will be nonvanishing. Here, $n_{i=1,2}$ labels single-atom eigenstates, E_{n_i} is the corresponding eigenenergy, and $|n_1, n_2\rangle$ represents the two-particle state with one atom in state n_1 and the other one in state n_2. Since the ground state $|0,0\rangle$ is isotropic, the summation is independent of the choice of Cartesian coordinates. The equation above can then be expressed as

$$
E^{(2)}(r) = \frac{6}{(4\pi\epsilon_0)^2 r^6} \sum_{(n_1,n_2)\neq(0,0)} \frac{|\langle n_1|d_{1x}|0\rangle|^2 |\langle n_2|d_{2x}|0\rangle|^2}{2E_0 - E_{n_1} - E_{n_2}}. \tag{5.3}
$$

We immediately see that this interaction potential varies as $1/r^6$, and is attractive since $E_{n_1} + E_{n_2} > 2E_0$. This $1/r^6$ attractive interaction is called the *van der Waals potential* and is the leading term in the long-distance part of the two-body interaction between two atoms in their ground states

$$V(r) = -\frac{C_6}{r^6} - \frac{C_8}{r^8} - \frac{C_{10}}{r^{10}} - \cdots .$$ (5.4)

The coefficient C_6 represents the strength of the van der Waals interaction and can be calculated from Eq. (5.3). The total interatomic interaction is then the combination of a short-distance repulsive core and a long-distance tail of the van der Waals potential, as illustrated in Fig. 5.1.

The values of C_6 for different alkali-metal atoms are listed in Table 5.1. Here, we have adopted as the length unit the Bohr radius a_0 and the energy units $\hbar^2/ma_0^2$. Notice that the coefficient for hydrogen atoms is almost three orders of magnitude smaller than other alkali-metal atoms. As a consequence, the s-wave scattering length between hydrogen atoms is also very small, leading to a small elastic collision rate. This fact makes

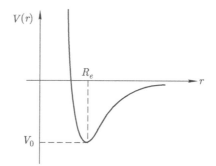

Figure 5.1.   A schematic drawing of interatomic interaction potential between two atoms in their ground states.

Table 5.1.   Characteristic values of van der Waals coefficients C_6 for several species of alkali-metal atoms.

Element	C_6
^{1}H	6.5 [2]
^{6}Li	1393 [3]
^{7}Li	1389 [2]
^{23}Na	1556 [2]
^{39}K	3897 [2]
^{40}K	3897 [4]
^{87}Rb	4698 [5]
^{133}Cs	6860 [6]

the evaporative cooling method much less efficient for hydrogen than for other alkali-metal atoms. Because of this, the Bose–Einstein condensation of atomic hydrogen was realized nearly three years after the first achievement of BEC [7], although the absolute temperature required for hydrogen is higher due to its light mass.

5.2.   Two-atom scattering in free space

Since the mutual distance between two atoms is much larger than the effective range of the interatomic interaction potential, the properties of the system are governed by the long-distance tail and are not sensitive to the short-distance details of the interatomic interaction potential. In quantum mechanics, the theory that discusses the long-wavelength, low-energy features of two-body physics is the scattering theory, which focuses mainly on the change between an incident and the scattered states when two atoms are apart. In this so-called *asymptotic* limit, the interaction potential between two atoms are negligible, such that the incident and scattered states are both eigenstates of the non-interacting Hamiltonian.

In this section, we consider a system of two distinguishable atoms, both with mass m interacting via an isotropic interatomic potential, where the Hamiltonian can be written as

$$H_0 = -\frac{\nabla_{\mathbf{r}_1}^2}{2m} - \frac{\nabla_{\mathbf{r}_2}^2}{2m} + V(|\mathbf{r}_1 - \mathbf{r}_2|). \tag{5.5}$$

Here, $\mathbf{r}_{i=1,2}$ labels the spatial coordinate of the i-th atom. By defining the center-of-mass (c.m.) coordinate $\mathbf{R} = (\mathbf{r}_1 + \mathbf{r}_2)/2$ and the relative coordinate $\mathbf{r} = \mathbf{r}_1 - \mathbf{r}_2$, the Hamiltonian becomes separable with a non-interacting c.m. part and an interacting relative part. The relative Hamiltonian takes the form

$$H_{\text{rel}} = -\frac{\nabla_{\mathbf{r}}^2}{2\mu} + V(r), \tag{5.6}$$

where $\mu \equiv m/2$ is the reduced mass. In this chapter, we discuss only the elastic process where the internal degrees of freedom of the atoms remain unchanged during scattering.

The relative wavefunction of the scattering states can be written in the following ansatz

$$\Psi(\mathbf{r}) = e^{ikz} + \Psi_{\text{sc}}(\mathbf{r}), \tag{5.7}$$

where the first term is the incident plane-wave state with energy $E = \hbar^2 k^2 / 2\mu$. Here, we assume without loss of generality that the atom is injected along the z-direction. The second term represents the scattered wavefunction, which becomes an outgoing spherical wave in the asymptotic limit

$$\Psi_{\mathrm{sc}}(\mathbf{r}) \to f(\mathbf{k}) \frac{e^{ikr}}{r}. \tag{5.8}$$

The function $f(\mathbf{k})$ is called the *scattering amplitude*, which by definition contains all information of the scattering process in the large interatomic separation limit. A central question in scattering theory involves obtaining $f(\mathbf{k})$.

Since the interacting potential $V(r)$ is spherically symmetric, the scattering wavefunction $\Psi(\mathbf{r})$ acquires cylindrical symmetry and the scattering amplitude $f(\mathbf{k})$ depends only on the scattering angle θ, which is defined as the angle between $\mathbf{k}$ and $\hat{z}$ (Fig. 5.2). It is then convenient to expand the wavefunction $\Psi(\mathbf{r})$ in terms of Legendre polynomials $P_\ell(\cos \theta)$

$$\Psi(\mathbf{r}) = \sum_{\ell=0}^{\infty} A_\ell P_\ell(\cos \theta) R_{k\ell}(r), \tag{5.9}$$

where the radial part $R_{k\ell}(r)$ satisfies the radial Schrödinger equation

$$R_{k\ell}''(r) + \frac{2}{r} R_{k\ell}'(r) + \left[k^2 - \frac{\ell(\ell+1)}{r^2} - \frac{2\mu}{\hbar^2} V(r) \right] R_{k\ell}(r) = 0. \tag{5.10}$$

In the large distance limit $r \to \infty$, the interaction potential $V(r)$ vanishes and the radial equation listed above can be solved analytically. This leads to an asymptotic form of the radial function

$$R_{k\ell}(r) \approx \frac{1}{kr} \sin \left(kr - \frac{\pi}{2} \ell + \delta_\ell \right), \tag{5.11}$$

where δ_ℓ is called the *phase shift*. It will soon be clear that the phase shift is of great importance in the scattering theory.

Figure 5.2.   Scattering process between two atoms.

By comparing Eqs. (5.7), (5.9) and (5.11), and expanding the plane wave e^{ikz} in the Legendre polynomials, we get the following forms for the expansion coefficients A_ℓ

$$A_\ell = i^\ell \, (2\ell + 1) \, e^{i\delta_\ell}, \tag{5.12}$$

and the scattering amplitude $f(\mathbf{k})$

$$f(\mathbf{k}) = \frac{1}{2ik} \sum_{\ell=0}^{\infty} (2\ell + 1) \left(e^{2i\delta_\ell} - 1 \right) P_\ell(\cos\theta). \tag{5.13}$$

Notice that the angular dependence of $f(\mathbf{k})$ is solely determined by the phase shift δ_ℓ. From the expression of scattering amplitude listed above, we can obtain the differential scattering cross section, which is defined as $d\sigma/d\Omega \equiv |f(\mathbf{k})|^2$. The total scattering cross section is obtained by integrating the differential cross section over all solid angles, leading to

$$\sigma \equiv \int d\Omega \frac{d\sigma}{d\Omega} = 2\pi \int_0^\pi |f(\mathbf{k})|^2 \sin\theta d\theta$$

$$= \frac{4\pi}{k^2} \sum_{\ell=0}^{\infty} (2\ell + 1) \sin^2 \delta_\ell. \tag{5.14}$$

Here, we have used the orthonormal property of the Legendre polynomials.

For a short-range interaction potential such as the van der Waals potential $V(r \to \infty) \sim 1/r^6$, it can be proven that the phase shifts for different values of ℓ satisfy [1]

$$\begin{cases} \delta_{\ell=0} \propto k, & s\text{-wave;} \\ \delta_{\ell=1} \propto k^3, & p\text{-wave;} \\ \delta_{\ell\geq2} \propto k^4, & d\text{- and higher partial waves.} \end{cases} \tag{5.15}$$

Thus, the total cross section σ is dominated by the $\ell = 0$, i.e., the s-wave term, for low-energy scattering processes. A simple explanation for this observation is that for higher partial waves, the scattering wavefunction hits an effective centrifugal barrier around $r = 0$. In the low-energy limit with $k \to 0$, the probability for the wavefunction to overcome this centrifugal barrier is negligible.

Considering only the s-wave contribution, the low-energy scattering amplitude becomes

$$f(\mathbf{k}) = \frac{1}{2ik}\left(e^{2i\delta_0} - 1\right) \approx \frac{\delta_0}{k}. \tag{5.16}$$

Notice that the scattering amplitude is now isotropic, as required by the s-wave nature of the scattering wavefunction. From this result, we can easily obtain the expressions for the total scattering cross section

$$\sigma \approx \frac{4\pi\delta_0^2}{k^2}, \tag{5.17}$$

and the asymptotic radial function of Eq. (5.11)

$$R_{k0}(r) \approx \frac{\sin kr}{kr}\cos\delta_0 + \frac{\cos kr}{kr}\sin\delta_0 \approx 1 + \frac{\delta_0}{k}\frac{1}{r}. \tag{5.18}$$

Notice that all interesting quantities are determined by the k-dependence of the s-wave phase shift δ_0. Specifically, if we define the limiting value of δ_0/k as

$$a_s \equiv -\lim_{k\to 0}\frac{\delta_0}{k}, \tag{5.19}$$

the low-energy scattering quantities will be solely described by a_s. The quantity a_s has the dimension of length and is known as the *s-wave scattering length*.

Before concluding this section, we consider the scattering of two atoms interacting via a three-dimensional square-well potential as a particular example:

$$V_{\mathrm{sq}}(r) = \begin{cases} V_0, & r < R \\ 0, & r > R \end{cases} \tag{5.20}$$

where $R > 0$ is the interaction range. The potential depth V_0 can either be negative or positive, corresponding to the case of attractive and repulsive interactions, respectively. Since the square-well potential is a short-range potential, the low-energy scattering is dominated by the s-wave portion, as we have discussed. The radial Schrödinger Eq. (5.10) for the s-wave case becomes

$$\left[\frac{d^2}{dr^2} - \frac{\mu V_{sq}(r)}{\hbar^2} + k^2 \right] u(r) = 0, \tag{5.21}$$

where we have defined $u(r) \equiv R_{k0}(r)r$. The general solutions of Eq. (5.21) can be written as

$$u_1(r) = A e^{ik_1 r} + B e^{-ik_1 r}, \quad \text{for } r < R,$$
$$u_2(r) = C e^{ikr} + D e^{-ikr}, \quad \text{for } r > R, \tag{5.22}$$

where $k_1 = \sqrt{k^2 - \mu V_0/\hbar^2}$.

To determine the coefficients, we need to exploit boundary conditions. First, the inner solution $u_1(r)$ must go to zero as $r \to 0$, otherwise the physical radial wavefunction $R_{k0}(r) = u_1(r)/r$ will diverge. Second, the outer solution should take the asymptotic form

$$u_2(r \to \infty) = e^{-ikr} - e^{2i\delta_0} e^{ikr}. \tag{5.23}$$

Finally, the inner and outer solutions need to be connected by the continuity condition at $r = R$. By substituting these boundary conditions into the general solutions (5.22), we obtain the expression for the phase shift

$$\delta_0 = -kR + \tan^{-1} \left[\frac{k}{k_1} \tan(k_1 R) \right]. \tag{5.24}$$

In the case of hard-core atoms with $V_0 \to +\infty$, we have $k_1 \to i\infty$ and $\delta_0 \to -kR$. Thus, the scattering length defined in Eq. (5.19) becomes $a_s = R$, which is just the radius of atoms. Due to this reason we include a minus sign in the definition of a_s in Eq. (5.19).

For the attractive case of $V_0 < 0$, the scattering length can be rewritten as

$$a_s = R \left(1 - \frac{\tan \gamma}{\gamma} \right), \tag{5.25}$$

where $\gamma = R\sqrt{\mu |V_0|/\hbar^2}$. The variation of a_s as a function of γ is shown in Fig. 5.3. Notice that a_s diverges and changes sign at positions of $\gamma = (n + 1/2)\pi$ with $n = 0, 1, 2, \ldots$. This diverging behavior of a_s is called a *potential resonance* or *shape resonance*. It happens each time as the potential is just deep enough to support a new bound state.

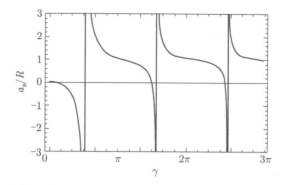

Figure 5.3.   Variation of s-wave scattering length as a function of $\gamma \equiv R\sqrt{\mu|V_0|/\hbar^2}$ of a three-dimensional attractive square potential. Note that a_s diverges and changes sign at $\gamma = (n+1/2)\pi$, where a shape resonance takes place.

5.3.   Effective interaction

In the previous section, we have shown that the low-energy scattering properties of two atoms interacting via a short-range potential can be described by a single quantity, i.e., the s-wave scattering length a_s. The physical meaning of this observation is quite clear: As the energy of the incident atom is small enough, the atom cannot penetrate into the core, but will be scattered away by the long-distance tail of the interacting potential. In this case, it is of no surprise that the short-distance details of the potential are irrelevant. To be specific, all physical quantities associated with the low-energy scattering process are functions of a_s only, e.g.,

$$f = -a_s, \tag{5.26a}$$

$$\sigma = 4\pi a_s^2, \tag{5.26b}$$

$$R_{00}(r) \approx 1 - \frac{a_s}{r}. \tag{5.26c}$$

One important application of this conclusion is to model the original, usually a very complicated interaction potential by a much simpler effective interaction potential when discussing physics in the low-energy, long-wavelength limit. In fact, we can easily conclude from the discussion above that if two interacting potentials give the same a_s, there is no way to distinguish them from low-energy scattering processes. In principle, one can use an arbitrary interaction potential to mimic the original Hamiltonian.

Here, we discuss three commonly used methods in cold atom physics, including the Bethe–Peierls boundary condition, the pseudopotential and the contact potential with renormalization. As we will see, these three methods have their own advantages in different situations and hence are most suitable when tackling the corresponding types of problem.

5.3.1. *Bethe–Peierls boundary condition*

In the study of cold atom gases, the Bethe–Peierls (BP) boundary condition is widely used as a replacement for the actual interaction between two atoms [8]. Within this approach, instead of employing a model interacting potential, we explicitly impose a boundary condition to guarantee that the wavefunction acquires the desired form of Eq. (5.26c) in the long-distance regime. As a consequence, we only need to solve the Schrödinger equation for the non-interacting Hamiltonian, and thus the calculation is usually significantly simplified. The BP boundary condition is very useful in the study of few-body and many-body physics in cold atom physics, especially those with large interatomic scattering lengths. For example, Petrov and co-workers obtained the atom-dimer [9] and the dimer-dimer [10] scattering lengths in two-component Fermi gases, while Werner and Castin [11] rederived the well-known Tan's relations [12] using the BP boundary condition.

Considering as an example the same two-atom scattering problem we have investigated in the previous section, the low-energy physics can be described by the free Hamiltonian $H = -\nabla_{\mathbf{r}}^2/2\mu$, together with the BP boundary condition

$$\Psi(r \to 0) \propto 1 - \frac{a_s}{r}. \tag{5.27}$$

It is then straightforward to show that the solution of this problem leads to Eq. (5.26).

The most appealing feature of the BP boundary condition method is its simplicity in calculation. In most cases, we can write down an analytic solution of the free Hamiltonian and substitute the expression into the BP condition. However, since this method requires an external input for the wavefunction's behavior, it is impractical to be implemented in problems with more than a few particles. In fact, even for the cases of three and four particles, the application of the BP boundary condition needs some special and sometimes complicated treatment [9, 10].

5.3.2. Huang–Yang pseudopotential

Among all the model potentials one can think of, the simplest candidate is the three-dimensional δ-potential $U\delta(\mathbf{r})$, with U the interaction strength. It is then natural to suggest using a δ-potential as an effective interaction to mimic the original Hamiltonian. However, this simple choice will introduce some trouble if one thinks a step further. Indeed, if we adopt the expression $V(\mathbf{r}) = U\delta(\mathbf{r})$ and set $k = \ell = 0$, the radial Eq. (5.10) becomes

$$R_{00}''(r) + \frac{2}{r}R_{00}'(r) - \frac{2\mu U}{\hbar^2}\delta(r)R_{00}(r) = 0. \tag{5.28}$$

By substituting the radial wavefunction Eq. (5.26c) $R_{00} = 1 - a_s/r$, we obtain the following formal equation[1]

$$4\pi a_s\delta(\mathbf{r}) - \frac{2\mu U}{\hbar^2}\delta(\mathbf{r})R_{00}(r) = 0. \tag{5.29}$$

Since the function $R_{00}(r)$ acquires a $1/r$ singularity at the origin, the equation listed above will never be satisfied at $r = 0$, regardless of the sign and value of the interacting strength U.

The divergence of the radial equation is an unphysical consequence of imposing the asymptotic form of wavefunction in the short-distance regime. In fact, the essence of effective interaction is to match the wavefunction of a desired form within the long-distance regime. Pushing the same form toward the short-distance core is unnecessary, and in some cases, like this one, give wrong results.

One method to overcome this issue is to introduce an s-wave pseudopotential,

$$V_{\mathbf{pp}}(r) = U_p\delta(\mathbf{r})\frac{\partial}{\partial r}(r \cdot). \tag{5.30}$$

The concept of pseudopotential was first discussed by Fermi [13] and generalized by Huang and Yang in their seminal paper in 1957 [14]. The differential operator $(\partial/\partial r)r$ in the pseudopotential may be replaced by unity if

---

[1]Here, we have used the relation

$$\nabla_{\mathbf{r}}^2\left(\frac{1}{r}\right) = -4\pi\delta(\mathbf{r}).$$

the subsequent wavefunction ψ is well behaved at $r = 0$, for then

$$\frac{\partial}{\partial r}(r\psi)\bigg|_{r=0} = \psi(0) + r\frac{\partial}{\partial r}\psi\bigg|_{r=0} = \psi(0). \tag{5.31}$$

If $\psi(r \to 0) = 1 - a_s/r$, however, the differential operator gives the non-trivial result of

$$\frac{\partial}{\partial r}(r\psi)\bigg|_{r=0} = 1. \tag{5.32}$$

Using the s-wave pseudopotential, the radial Eq. (5.10) can be reduced to the following form

$$4\pi a_s\delta(\mathbf{r}) - \frac{2\mu}{\hbar^2}\delta(\mathbf{r}) = 0, \tag{5.33}$$

which leads to the result

$$U_p = \frac{4\pi\hbar^2 a_s}{m}. \tag{5.34}$$

Here, we change the notation back to the atomic mass $m = 2\mu$ to make the expression consistent with general literature.

We note that the pseudopotential Eq. (5.30) is not a Hermitian operator because the derivative $(\partial/\partial r)r$ is not Hermitian. The non-Hermiticity reflects the fact that the wavefunctions of this effective Hamiltonian do not coincide everywhere with the eigenfunctions of the original system, but do so only in the asymptotic region. The eigenvalues of the pseudopotential, however, are all real numbers and are the approximate solutions of the real problem. One thing we should bear in mind is that this non-Hermitian Hamiltonian does not qualify the Reyleigh–Ritz variational theorem, hence cannot be directly solved via a conventional variational method.

5.3.3.   Contact potential with renormalization

In addition to the pseudopotential, an alternative method to bypass the divergence behavior of a δ-potential is renormalization. While the pseudopotential works in the spatial coordinate and removes the singularity at a short-distance limit, the renormalization procedure is carried out in momentum space and regularizes the high-energy degrees of freedom therein.

The connection between these two methods can be clearly seen by noticing that we obtain a constant interaction strength for arbitrary momenta by Fourier transforming the δ potential into momentum space. This unphysical result introduces an ultraviolet divergence contributed by the large-momentum degrees of freedom. Within the pseudopotential approach, the derivative operator $(\partial/\partial r)r$ removes the singularity of the wavefunction and leads to a constant value at $r \to 0$. This procedure hence sets an effective range of the wavefunction in momentum space, which in turn regularizes the ultraviolet divergence. Bearing the same spirit, the renormalization method works directly in momentum space.

To see how the renormalization works, we first revisit the two-atom scattering problem in free space, but this time in momentum space. The relative wavefunction Eq. (5.7) then becomes

$$\Psi(\mathbf{k}_1) = (2\pi)^3 \delta(\mathbf{k}_1 - \mathbf{k}) + \Psi_{\mathrm{sc}}(\mathbf{k}_1). \tag{5.35}$$

This wavefunction satisfies the Schrödinger equation

$$\left(\frac{\hbar^2 k^2}{2\mu} - \frac{\hbar^2 k_1^2}{2\mu}\right) \Psi_{\mathrm{sc}}(\mathbf{k}_1) = V(\mathbf{k}_1, \mathbf{k}) + \frac{1}{\mathcal{V}} \sum_{\mathbf{k}_2} V(\mathbf{k}_1, \mathbf{k}_2) \Psi_{\mathrm{sc}}(\mathbf{k}_2), \tag{5.36}$$

where the scattering energy $E = \hbar^2 k^2/2\mu$, $\mathcal{V}$ is the quantization volume and $V(\mathbf{k}_1, \mathbf{k})$ is the Fourier transform of the interaction potential

$$V(\mathbf{k}_1, \mathbf{k}_2) = V(\mathbf{k}_1 - \mathbf{k}_2) = \int d^3 \mathbf{r} e^{-i(\mathbf{k}_1 - \mathbf{k}_2) \cdot \mathbf{r}} V(\mathbf{r}). \tag{5.37}$$

The formal solution of Eq. (5.36) can be easily obtained as

$$\Psi_{\mathrm{sc}}(\mathbf{k}_1) = \left(\frac{\hbar^2 k^2}{2\mu} - \frac{\hbar^2 k_1^2}{2\mu} + i\delta\right)^{-1} \left[V(\mathbf{k}_1, \mathbf{k}) + \frac{1}{\mathcal{V}} \sum_{\mathbf{k}_2} V(\mathbf{k}_1, \mathbf{k}_2) \Psi_{\mathrm{sc}}(\mathbf{k}_2)\right]$$

$$\equiv \left(\frac{\hbar^2 k^2}{2\mu} - \frac{\hbar^2 k_1^2}{2\mu} + i\delta\right)^{-1} T(\mathbf{k}_1, \mathbf{k}; E). \tag{5.38}$$

Here, an infinitesimal imaginary part δ has been included to ensure that only outgoing waves are present in the scattered wavefunction. In the second line of Eq. (5.38), we define a *scattering matrix* $T(\mathbf{k}_1, \mathbf{k}; E)$ (also referred to as a *transition matrix* or *T-matrix*), which satisfies the so-called

Lippmann–Schwinger equation [15]

$$T(\mathbf{k}_1, \mathbf{k}; E) = V(\mathbf{k}_1, \mathbf{k}) + \frac{1}{\mathcal{V}} \sum_{\mathbf{k}_2} V(\mathbf{k}_1, \mathbf{k}_2) \left(E - \frac{\hbar^2 k_2^2}{2\mu} + i\delta \right)^{-1} T(\mathbf{k}_2, \mathbf{k}; E).$$
$$(5.39)$$

In the long-wavelength and low-energy limit, we set $E = k = 0$ and rewrite Eq. (5.38) as

$$\Psi_{\rm sc}(\mathbf{k}_1) = \left(-\frac{\hbar^2 k_1^2}{2\mu} + i\delta \right)^{-1} T(\mathbf{k}_1, 0; 0). \qquad (5.40)$$

By Fourier transforming back to real space, we obtain

$$\Psi_{\rm sc}(\mathbf{r}) = \int \frac{d^3 \mathbf{k}_1}{(2\pi)^3} e^{i\mathbf{k}_1 \cdot \mathbf{r}} \Psi_{\rm sc}(\mathbf{k}_1)$$
$$= \int \frac{d^3 \mathbf{k}_1}{(2\pi)^3} \frac{e^{i\mathbf{k}_1 \cdot \mathbf{r}}}{-\hbar^2 k_1^2/2\mu + i\delta} T(\mathbf{k}_1, 0; 0). \qquad (5.41)$$

The asymptotic behavior of this wavefunction can be extracted by noticing that in the long-distance limit $r \to \infty$, the large-momentum contribution to the integral over $d^3 \mathbf{k}_1$ is negligible due to the fast oscillation of the exponential factor $e^{i\mathbf{k}_1 \cdot \mathbf{r}}$, and the integral is dominated by the neighborhood around $k_1 \approx 0$. Thus, we have

$$\Psi_{\rm sc}(\mathbf{r}) \approx -\frac{2\mu T(0,0;0)}{\hbar^2} \int \frac{d^3 \mathbf{k}_1}{(2\pi)^3} \frac{e^{i\mathbf{k}_1 \cdot \mathbf{r}}}{k_1^2}$$
$$= -\frac{2\mu}{4\pi\hbar^2 r} T(0,0;0). \qquad (5.42)$$

Here, we have used the identity

$$\int \frac{d^3 \mathbf{k}_1}{(2\pi)^3} \frac{e^{i\mathbf{k}_1 \cdot \mathbf{r}}}{k_1^2} = \frac{1}{4\pi r}. \qquad (5.43)$$

Comparing the result (5.42) with Eq. (5.26c), we obtain the expression for T-matrix in the low-energy limit

$$T(0,0;0) = \frac{4\pi\hbar^2 a_s}{m}, \qquad (5.44)$$

where $m = 2\mu$ is the atomic mass. More generally, the T-matrix is related to the scattering amplitude via [1]

$$T(\mathbf{k}_1, \mathbf{k}; E = \hbar^2 k^2 / m) = -\frac{4\pi\hbar^2}{m} f(\mathbf{k}, \mathbf{k}_1). \qquad (5.45)$$

From the discussion above, we show that the low-energy, long-distance scattering physics is now described by the zero-energy T-matrix $T(0, 0; 0)$, which is proportional to the s-wave scattering length a_s. An effective interaction can successfully mimic the low-energy physics of the real problem provided that it can give the very same T-matrix.

Now we proceed to investigate the contact potential $V(\mathbf{r}) = U\delta(\mathbf{r})$. By transforming the potential to momentum space, we obtain the Lippmann–Schwinger equation for the δ-potential

$$T(\mathbf{k}_1, \mathbf{k}; E) = U + U \sum_{\mathbf{k}_2} \left(E - \frac{\hbar^2 k_2^2}{2\mu} + i\delta \right)^{-1} T(\mathbf{k}_2, \mathbf{k}; E). \qquad (5.46)$$

By multiplying $(E - \hbar^2 k_1^2 / 2\mu + i\delta)^{-1}$ on both sides of Eq. (5.46) and summing over $\mathbf{k}_1$, we get

$$\sum_{\mathbf{k}_1} \left(E - \frac{\hbar^2 k_1^2}{2\mu} + i\delta \right)^{-1} T(\mathbf{k}_1, \mathbf{k}; E) = U \sum_{\mathbf{k}_1} \left(E - \frac{\hbar^2 k_1^2}{2\mu} + i\delta \right)^{-1}$$

$$+ U \sum_{\mathbf{k}_1} \left(E - \frac{\hbar^2 k_1^2}{2\mu} + i\delta \right)^{-1} \sum_{\mathbf{k}_2} \left(E - \frac{\hbar^2 k_2^2}{2\mu} + i\delta \right)^{-1} T(\mathbf{k}_2, \mathbf{k}; E).$$

$$(5.47)$$

This equation can be rewritten into another form

$$\sum_{\mathbf{k}_1} \left(E - \frac{\hbar^2 k_1^2}{2\mu} + i\delta \right)^{-1}$$

$$\times \left\{ T(\mathbf{k}_1, \mathbf{k}; E) \left[1 - U \sum_{\mathbf{k}_2} \left(E - \frac{\hbar^2 k_2^2}{2\mu} + i\delta \right)^{-1} \right] - U \right\} = 0. \qquad (5.48)$$

Here, we have interchanged dummy variables $\mathbf{k}_1 \leftrightarrow \mathbf{k}_2$ in Eq. (5.47).

In the low-energy limit with $E \to 0$ and $k \to 0$, the summation over $\mathbf{k}_1$ in the expression above is dominated by the term with $\mathbf{k}_1 = 0$.

Thus, we have

$$T(0,0;0)\left(1 + U\sum_{\mathbf{k}}\frac{1}{2\epsilon_{\mathbf{k}}}\right) - U = 0, \qquad (5.49)$$

where $\epsilon_{\mathbf{k}} = \hbar^2 k^2/2m$ is the free particle dispersion. By substituting the desired form of Eq. (5.44) for the low-energy T-matrix into Eq. (5.49), we obtain the expression for the contact potential strength

$$\frac{1}{U} = \frac{m}{4\pi\hbar^2 a_s} - \sum_{\mathbf{k}}\frac{1}{2\epsilon_{\mathbf{k}}}. \qquad (5.50)$$

A careful reader may immediately feel uncomfortable about the summation over $\mathbf{k}$ in Eq. (5.50). This term obviously diverges in the ultraviolet regime, thus the interaction strength U is meaningless at first glance. This uneasy feeling, however, is in some sense a correct understanding of the renormalization procedure. As we have already learned, the contact potential is by all means an ill-behaved potential and the application of this effective model will inevitably cause some unphysical consequences. In other words, there is no way we can find a "meaningful" contact potential that can reproduce the low-energy scattering physics as the original system. The expression of Eq. (5.50) defines an artificial contact potential with an obvious divergence in the interaction strength. However, this divergence associated with U will cancel the divergence introduced by the δ-function of the contact potential. As a result, we can reproduce the correct s-wave scattering length a_s, which guarantees the validity of the effective interaction. In this sense, the "meaningless" U is usually referred to as the *bare interaction*, while the well-defined $U_p \equiv 4\pi\hbar^2 a_s/m$ is called the *physical interaction*.

In contrast to the BP boundary condition and pseudopotential methods, the renormalization procedure directly deals with a contact interaction in momentum space and hence it is natural for it to be implemented in the many-body analysis. The expression of Eq. (5.50) will be used intensively in the remainder of this book.

References

[1] J. R. Taylor. *Scattering Theory: The Quantum Theory of Nonrelativistic Collisions*. John Wiley & Sons, New York (1972).
[2] A. Derevianko, J. F. Babb and A. Dalgarno. *Phys. Rev. A* **63**, 052704 (2001).

[3] Z.-C. Yan, J. F. Babb, A. Dalgarno and G. W. F. Drake. *Phys. Rev. A* **54**, 2824 (1996).

[4] A. Derevianko, W. R. Johnson, M. S. Safronova and J. F. Babb. *Phys. Rev. Lett.* **82**, 3589 (1999).

[5] E. G. M. van Kempen, S. J. J. M. F. Kokkelmans, D. J. Heinzen and B. J. Verhaar. *Phys. Rev. Lett.* **88**, 093201 (2002).

[6] C. Chin and V. Vulentić, A. J. Kerman, S. Chu, E. Tiesinga, P. J. Leo and C. J. Williams. *Phys. Rev. A* **70**, 032701 (2004).

[7] D. G. Fried, T. C. Killian, L. Willmann, D. Landhuis, S. C. Moss, D. Kleppner and T. J. Greytak. *Phys. Rev. Lett.* **81**, 3811 (1998).

[8] H. Bethe and R. Peierls. *Proc. R. Soc. A* **148**, 146 (1935).

[9] D. S. Petrov. *Phys. Rev. A* **67**, 010703 (2003).

[10] D. S. Petrov, C. Salomon and G. V. Shlyapnikov. *Phys. Rev. Lett.* **93**, 090404 (2004).

[11] F. Werner and Y. Castin. *Phys. Rev. A* **86**, 013626 (2012).

[12] S. Tan. *Ann. Phys.* **323**, 2952 (2008); *ibid.*, 2971 (2008); *ibid.*, 2987 (2008).

[13] E. Fermi. La Ricerca Seicntifica VII-II, 13 (1936).

[14] K. Huang and C. N. Yang. *Phys. Rev.* **105**, 767 (1957).

[15] B. A. Lippmann and J. Schwinger. *Phys. Rev. Lett.* **79**, 469 (1950).

6
Feshbach Resonance

In the previous chapter, we have learned that the long-wavelength, low-energy scattering process between two distinguishable atoms is determined by the s-wave scattering length a_s. This single-value parameter thus governs the effective interaction strength between the two atoms via, e.g., the renormalization relation Eq. (5.50). It is then of great interest if one can find a way to tune a_s, especially toward the strongly interacting regime.

One way to change a_s is by tuning through a shape resonance, as discussed in Sec. 5.2. This can be achieved by either increasing the depth or varying the shape of the interatomic interaction curve. As a particular example of a three-dimensional square-well potential with radius R, the s-wave scattering length varies continuously between negative infinity and positive infinity with increasing interaction-potential depth V_0, and changes sign at the resonance position of $V_0 = -(n+1/2)^2 \pi^2 \hbar^2 / \mu R^2$, where a new bound state emerges (see Fig. 5.3). However, this method requires a controllable, and usually significant modification of the interaction potential between two atoms, which is not an easy task in experiments.

An alternative (and more applicable) approach to tune a_s is to employ a mechanism called *Feshbach resonance*, or equivalently the *Fano–Feshbach resonance*, named after Ugo Fano [1] and Herman Feshbach [2, 3]. In their works, Fano and Feshbach independently investigated the resonance phenomena that arise from the coupling of a discrete state to the continuum threshold. While Fano developed his treatment in the context of atomic physics, Feshbach tackled the problem on the background of nuclear physics. In cold atom physics, Feshbach resonances were first considered by Stwalley [4] and Reynold *et al.* [5] as an effect one should avoid to prevent

severe inelastic atom loss. The positive aspect of the Feshbach resonance
was first discussed in 1993 by Tiesinga *et al.* [6], who pointed out that
the sign and value of the *s*-wave scattering length between ultracold atoms
could be changed by tuning through the resonance point. This remarkable
idea was soon demonstrated by Inouye *et al.* [7] in ^{23}Na BEC and Courteille
et al. [8] in trapped ^{85}Rb atoms.

Nowadays, the Feshbach resonance is one of the most important techniques in cold atom experiments. For Bose gases, one needs to tune the resonance to obtain a desirable value of a_s, either to obtain large condensates
(^{7}Li [9, 10]) or to achieve BEC (^{85}Rb [11], ^{133}Cs [12, 13], ^{39}K [14]). For
Fermi systems, Feshbach resonances serve as an essential ingredient to
explore many-body physics in the strongly interacting regime. This regime
is realized when the scattering length a_s is tuned to exceed the interatomic
spacing. The first Feshbach resonance in a two-component Fermi gas was
observed by Loftus *et al.* [15] and was used to create the first strongly
interacting Fermi gas by O'Hara *et al.* [16]. In 2003, the Bose–Einstein condensation of molecules was created in atomic Fermi gases of ^{6}Li [17–19] and
^{40}K [20]. The most remarkable achievement along this line is the realization
of the BCS-BEC crossover [21–26], which has been theoretically studied in
the condensed matter community since 1980 [27] but is considered to be
impractical in any sense. In addition to the *s*-wave Feshbach resonance
between two distinguishable fermionic atoms at different hyperfine states,
p-wave Feshbach resonance between two fermions of the same hyperfine
state was also studied [28–30].

In this chapter, we first introduce the basic mechanism and some fundamental properties of the Feshbach resonance in Sec. 6.1. Feshbach resonance is a general phenomenon which can be realized in different systems
via various schemes. Thus, we intentionally constrain ourselves within a
phenomenological level and use an effective two-channel model to give an
outline of the underlying physics. Specific experimental schemes to realize
a Feshbach resonance in cold atom systems are then discussed in Secs. 6.2
and 6.3, where magnetically and optically tunable Feshbach resonances are
introduced, respectively.

6.1.   Basic physics of the Feshbach resonance

A scattering resonance between two atoms can take place as an isolated
bound state is brought close to the threshold of a continuum spectrum,

such that two incident atoms with threshold energy can be scattered to the bound state and subsequently decay back to the continuum threshold. Through this process, the effective interaction between the two atoms can be tuned by either changing the detuning or the coupling between the bound state and the threshold. In the case of shape resonance, both the continuum threshold and the isolated bound state are supported by the same interaction potential. Thus, the only way to vary their relative distance is to change the interaction itself.

The essence of the Feshbach resonance is to exploit the multi-channel nature of the atomic scattering process, which is rooted in the fact that atoms do have internal degrees of freedom instead of being point particles. Indeed, for each complete set of atomic quantum numbers, usually referred to as a *scattering channel*, there is a distinct interaction-potential curve in general. Thus, if a bound state associated with one channel is brought nearly degenerate to the continuum threshold of another channel, and the matrix elements coupling these two channels are nonzero, a resonance can take place between two atoms scattered in the threshold channel. Since the bound state and the threshold correspond to different channels, it is much easier to experimentally tune the detuning between them.

Needless to say, the physics around a Feshbach resonance between two atoms can be described beautifully if we employ a multi-channel calculation based on realistic interaction potentials. Much effort has been devoted along this line to account for a variety of experimental collisional and bound-state properties of alkali-metal species [15, 31–36]. However, this type of calculation is not readily accessible due to its complicated nature and the close dependence on the details of interaction potentials. As an alternative, the physics of the two-atom scattering process in the close vicinity of a resonance point can also be well described by an effective model that incorporates only the two nearly resonant channels [37–40]. Next, we adopt an effective two-channel model to investigate the properties of a Feshbach resonance. We denote the channel of continuum threshold as the *open channel* (o) and the channel of bound state as the *closed channel* (c). The general Hamiltonian for the relative motion of an atom pair is then given by the following matrix form

$$H_{2c} = \begin{pmatrix} H_{oo} & H_{oc} \\ H_{oc}^{\dagger} & H_{cc} \end{pmatrix}.$$ (6.1)

While the diagonal elements H_{oo} and H_{cc} correspond to the Hamiltonian projection into the subspaces associated with the open and closed channels, respectively, the off-diagonal terms represent the inter-channel coupling. An arbitrary quantum state can be written as a linear combination of its corresponding projections in the two subspaces

$$|\Psi\rangle = |\Psi_o\rangle + |\Psi_c\rangle, \tag{6.2}$$

where

$$|\Psi_o\rangle = \hat{P}_o|\Psi\rangle, \quad |\Psi_c\rangle = \hat{P}_c|\Psi\rangle. \tag{6.3}$$

Here, we have defined projection operators $\hat{P}_o$ and $\hat{P}_c$ for the open and closed subspaces, respectively, which satisfy the relations

$$\hat{P}_o^2 = \hat{P}_o, \quad \hat{P}_c^2 = \hat{P}_c,$$
$$\hat{P}_o + \hat{P}_c = 1, \quad \hat{P}_o\hat{P}_c = \hat{P}_c\hat{P}_o = 0. \tag{6.4}$$

By substituting Eq. (6.2) into the Schrödinger equation and applying the projection operators to the left, we obtain

$$\hat{P}_o H \left(|\Psi_o\rangle + |\Psi_c\rangle \right) = \hat{P}_o E \left(|\Psi_o\rangle + |\Psi_c\rangle \right) = E|\Psi_o\rangle,$$
$$\hat{P}_c H \left(|\Psi_o\rangle + |\Psi_c\rangle \right) = \hat{P}_c E \left(|\Psi_o\rangle + |\Psi_c\rangle \right) = E|\Psi_c\rangle. \tag{6.5}$$

These equations lead to

$$(E - H_{oo})|\Psi_o\rangle = H_{oc}|\Psi_c\rangle, \tag{6.6a}$$

$$(E - H_{cc})|\Psi_c\rangle = H_{oc}^{\dagger}|\Psi_o\rangle. \tag{6.6b}$$

We next substitute the formal solution of Eq. (6.6b)

$$|\Psi_c\rangle = \frac{1}{E - H_{cc} + i\delta} H_{oc}^{\dagger}|\Psi_o\rangle \tag{6.7}$$

into Eq. (6.6a) and obtain

$$(H_{oo} + H_{oo}')|\Psi_o\rangle = E|\Psi_o\rangle, \tag{6.8}$$

where

$$H_{oo}' \equiv \frac{H_{oc} H_{oc}^{\dagger}}{E - H_{cc} + i\delta} \tag{6.9}$$

acts as an induced interaction within the open channel. From this expression, we can see that if the energy E is nearly resonant with a bound state in the closed channel $E \approx H_{cc}$, the induced interaction in the open channel will drive the system toward resonance.

To illustrate this formal theory more clearly, we consider an example of two atoms interacting via three-dimensional square-well potentials in both the open and closed channels. The elements in the Hamiltonian Eq. (6.1) are

$$H_{oo} = -\frac{\hbar^2 \nabla_{\mathbf{r}}^2}{2\mu} + V_o(\mathbf{r}),$$

$$H_{cc} = -\frac{\hbar^2 \nabla_{\mathbf{r}}^2}{2\mu} + V_c(\mathbf{r}) + \nu,$$

$$H_{oc} = H_{oc}^\dagger = W(\mathbf{r}). \tag{6.10}$$

Here, we assume the open and closed channels are shifted by an energy difference ν and the inter-channel coupling is a real function without loss of generality. For simplicity, we consider the case where $V_o(\mathbf{r})$, $V_c(\mathbf{r})$ and $W(\mathbf{r})$ acquire the same range, as can be expressed in the following form

$$V_{c,o}(\mathbf{r}), W(\mathbf{r}) = \begin{cases} -V_{c,o}, W & \text{for } r < R, \\ 0, & \text{for } r > R \end{cases} \tag{6.11}$$

with $V_o, V_c > 0$. A schematic drawing of the potentials is shown in Fig. 6.1. The Hamiltonian thus takes the form

$$H_{2c}^> = \begin{pmatrix} \frac{\hbar^2 \nabla_{\mathbf{r}}^2}{2\mu} & 0 \\ 0 & -\frac{\hbar^2 \nabla_{\mathbf{r}}^2}{2\mu} + \nu \end{pmatrix} \tag{6.12}$$

for $r > R$, and becomes

$$H_{2c}^< = \begin{pmatrix} -\frac{\hbar^2 \nabla_{\mathbf{r}}^2}{2\mu} - V_o & W \\ W & -\frac{\hbar^2 \nabla_{\mathbf{r}}^2}{2\mu} - V_c + \nu \end{pmatrix} \tag{6.13}$$

for $r < R$.

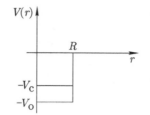

Figure 6.1.   Square-well potential curves in the open and closed channels.

To obtain the s-wave scattering length within the open channel, we need to solve for the scattering wavefunction in the low-energy limit $E = \hbar^2 k^2/2\mu \to 0$. Since the Hamiltonian in the long-distance regime $H_{2c}^{>}$ is diagonal, the radial wavefunctions $u_{o,c}^{>}(r) \equiv r R_{o,c}^{>}(r)$ within this region can be expressed as

$$\begin{pmatrix} u_o^{>} \\ u_c^{>} \end{pmatrix} = \begin{pmatrix} C e^{ikr} + D e^{-ikr} \\ F e^{-\kappa_c r} \end{pmatrix}, \tag{6.14}$$

where $k = \sqrt{2\mu E/\hbar^2}$ and $\kappa_c = \sqrt{2\mu\nu/\hbar^2 - k^2}$. In the region $r < R$ the solutions are of the form

$$\begin{pmatrix} u_o^{<} \\ u_c^{<} \end{pmatrix} = \begin{pmatrix} A_1 e^{ik_o^{<}r} - A_2 e^{-ik_o^{<}r} \\ B_1 e^{ik_c^{<}r} - B_2 e^{-ik_c^{<}r} \end{pmatrix}, \tag{6.15}$$

where

$$k_o^{<} = \sqrt{\frac{-2\mu\epsilon_o}{\hbar^2} + k^2},$$

$$k_c^{<} = \sqrt{\frac{-2\mu\epsilon_c}{\hbar^2} + k^2}. \tag{6.16}$$

Here, the energies ϵ_o and ϵ_c

$$\epsilon_o = \frac{-V_o - V_c + \nu}{2} - \frac{1}{2}\sqrt{(V_o - V_c + \nu)^2 + 4W^2},$$

$$\epsilon_c = \frac{-V_o - V_c + \nu}{2} + \frac{1}{2}\sqrt{(V_o - V_c + \nu)^2 + 4W^2} \tag{6.17}$$

are the eigenvalues of the matrix

$$V_{2c}^< = \begin{pmatrix} -V_o & W \\ W & -V_c + \nu \end{pmatrix}. \tag{6.18}$$

The coefficients in Eqs. (6.14) and (6.15) can be determined by matching boundary conditions. First, the radial wavefunctions $R_{o,c}^<$ must be regular as $r \to 0$. This condition requires $A_1 = A_2 = A$ and $B_1 = B_2 = B$. Second, the outer solution of the open channel should take the asymptotic form

$$u_o^>(r \to \infty) = e^{-ikr} - e^{2i\delta_0} e^{ikr} \tag{6.19}$$

with δ_0 being the s-wave phase shift. Finally, the inner and outer solutions and their first derivatives need to be connected at $r = R$ by continuity conditions. As a result, the phase shift takes the following form

$$\delta_0 = -k \left[R - \frac{\tan(k_o^< R)}{k_o^<} \right]. \tag{6.20}$$

The s-wave scattering length thus becomes

$$a_s = - \lim_{k \to 0} \frac{\delta_0}{k} = R \left[1 - \frac{\tan(k_o^< R)}{k_o^< R} \right]. \tag{6.21}$$

Notice that the expression above takes the very same form as Eq. (5.25), reflecting a resonance behavior of a_s at the positions of $k_o^< R = (n + 1/2)\pi$. However, unlike the case of shape resonance where a_s can only be tuned via potential depth and range, $k_o^< R$ can this time also be varied by changing the energy shift ν between the open and closed channels, as shown in Fig. 6.2. Specifically, if the energy shift is large enough with $\nu \to \infty$, we have $\epsilon_o \to -V_o$ and

$$a_s \to a_{bg} = R \left[1 - \frac{\hbar}{R\sqrt{2\mu V_o}} \tan\left(\frac{R\sqrt{2\mu V_o}}{\hbar} \right) \right]. \tag{6.22}$$

This far-off-resonant value of a_s is referred to as the *background scattering length*. As we move from this far-detuned position toward a resonance point, the scattering length is characterized by the near-resonant expression

$$a_s(\nu) = a_{bg} \left(1 - \frac{\Delta\nu}{\nu - \nu_0} \right), \tag{6.23}$$

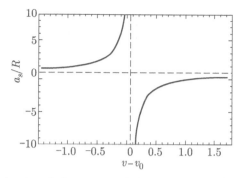

Figure 6.2.   Scattering length for two coupled square-well potentials as a function of the energy shift ν of the closed channel relative to the open channel. Using the length unit of R and energy unit of $\hbar^2/(2\mu R^2)$, the depths of the open and closed channels are $V_\text{o} = 5$ and $V_\text{o} = 3$, respectively, and the inter-channel coupling is $W = 1$. The horizontal dashed line indicates the background scattering length a_bg.

where ν_0 labels the resonance position and $\Delta\nu$ represents the width of the resonance. This result explicitly shows that the scattering length, and therefore the magnitude of the effective interatomic interaction, can be tuned to any value by sweeping the energy shift ν between the open and closed channels.

Next, we calculate the energy of the molecular state, which corresponds to the solution with the negative energy $-|E_b|$ of the Hamiltonian Eqs. (6.12) and (6.13). In this case, the radial wavefunctions in the region of $r > R$ become

$$\begin{pmatrix} u_\text{o}^> \\ u_\text{c}^> \end{pmatrix} = \begin{pmatrix} Ce^{-\kappa_\text{o} r} \\ Fe^{-\kappa_\text{c} r} \end{pmatrix}, \tag{6.24}$$

where $\kappa_\text{o} = \sqrt{2\mu|E_b|/\hbar^2}$ and $\kappa_\text{c} = \sqrt{2\mu(\nu + |E_b|)/\hbar^2}$. The wavefunctions for the inner region remain the same form as in Eq. (6.15), but with

$$k_\text{o}^< = \sqrt{\frac{-2\mu(\epsilon_\text{o} - |E_b|)}{\hbar^2}},$$

$$k_\text{c}^< = \sqrt{\frac{-2\mu(\epsilon_\text{c} - |E_b|)}{\hbar^2}}. \tag{6.25}$$

By joining the inner and outer solutions at $r = R$, we obtain the condition for a bound state with binding energy $|E_b|$ to exist in the open channel

$$\frac{-1}{\kappa_o} = \frac{\tan{(k_o^< R)}}{k_o^<}. \tag{6.26}$$

It is easy to verify that in the vicinity of a resonance point ν_0, a bound state does not exist for positive energy shift $\nu - \nu_0 > 0$. Instead, it starts to emerge at resonance and the binding energy can be expressed as

$$-|E_b| = -\frac{\hbar^2}{2\mu a_s^2(\nu)} \tag{6.27}$$

for $\nu - \nu_0 < 0$. A typical example of the variation of $-|E_b|$ near a Feshbach resonance is shown in Fig. 6.3.

Thus far, we have introduced the mechanism and some basic properties of a Feshbach resonance based on an effective two-channel model. The discussion is carried intentionally within a phenomenological level, in order to give a general description that is valid for different experimental scenarios. In the rest of this chapter, we will discuss various existing schemes to realize a Feshbach resonance in cold atom systems. For each case, the physical meanings of the two interacting channels, the inter-channel coupling and the relative energy shift will be clear.

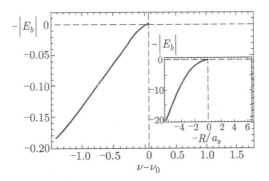

Figure 6.3.   Bound-state energy near a Feshbach resonance for two coupled square-well potentials as a function of ν. The inset shows the same result as a function of s-wave scattering length. Parameters used here are the same as in Fig. 6.2.

6.2.   Magnetic Feshbach resonance

As the most popular choice in present experiments, magnetically tunable Feshbach resonance has the advantages of easy implementation, wide tunability, small atom-loss rate and affordable heating effect. In this scheme, the two channels are related to different atomic hyperfine states, the interchannel coupling is induced by interatomic interaction potentials and the energy shift is the Zeeman splitting, which can be adjusted by tuning an external magnetic field.

The origin of a magnetic Feshbach resonance is rooted in the fact that the interatomic interactions between two neutral atoms are multi-channel in nature, depending on the internal degrees of freedom of the colliding atoms. For the most relevant case of alkali-metal atoms, only one electron is present in the outer shell. The unpaired electronic spins s_1 and s_2 from each atom can be coupled to a total spin $\mathbf{S} = \mathbf{s}_1 + \mathbf{s}_2$ with two possible choices $S = 0$ or 1. States with $S = 0$ and $S = 1$ are called singlet and triplet states, respectively. The interatomic interaction potentials are different for the singlet and triplet states.

In Fig. 6.4, the potential curves associated with the singlet $V_S(r)$ and triplet $V_T(r)$ states for two Li atoms are shown as an example. Similar potentials exist for any pair of alkali-metal atoms, either of the same or different species. For the case of two s-wave atoms, the long-distance tail of the two interaction potentials are identical, both of the van der Waals form with the leading term $-C_6/r^6$ as in Eq. (5.4). The splitting between the singlet and triplet potentials is from their difference in chemical bonding interactions when electronic clouds of the two atoms overlap at small interatomic distances $r \sim a_0$. The long-range form of this splitting is associated with the electron exchange interaction, which decreases exponentially with increasing r.

As we have learned in Sec. 2.3, the single-atom eigenstates at a zero magnetic field are the hyperfine states labeled by the total angular momentum F and its magnetic quantum number m_F. States with different values of F are separated by the hyperfine coupling $2\gamma_{\text{hfs}}$. For the same value of F, states with different m_F are degenerate. This degeneracy is lifted in the presence of a finite magnetic field, as shown in Fig. 2.3. Thus, as the two colliding atoms are separated far enough, such a two-body system can be characterized by the following set of quantum numbers

$$\alpha = \{F_1, m_{F_1}, F_2, m_{F_2}, \ell, m_\ell\}, \qquad (6.28)$$

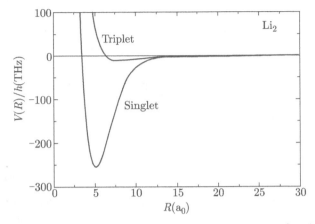

Figure 6.4. The singlet and triplet interacting potential curves for the Li$_2$ dimer molecule composed of two separated $^1S_{1/2}$ Li atoms. From Ref. [41].

where F_i and m_{F_i} label the single atomic hyperfine state of the i-th atom, and ℓ and m_ℓ represent the orbital state of the relative motion.[1]

Each set of the quantum number combination α is referred to as a channel. The channels with threshold energy $E_\alpha = E_{F_1, m_{F_1}} + E_{F_2, m_{F_2}}$ less than the total energy of the system E are considered as open channels. Conversely, those with energy $E_\alpha > E$ are called closed channels. If the system energy E is lower than the lowest energy channel, all channels are closed and the system can only be in one of the multiple isolated molecular levels. If E is higher than the lowest energy channel, then at least one channel is open and the system can be in the scattering continuum of the open channels.

Consider the simple case where only one channel labeled with α is open. We define the energy E_α as the energy zero point, such that the low-energy s-wave scattering process corresponds to the limit $E \rightarrow 0$. It is therefore convenient to represent the wavefunctions for the scattering or

---

[1]In principle, in the limit of large magnetic field strength, the Zeeman energy of the unpaired electron becomes comparable to the hyperfine coupling γ_{hfs}, and the total angular momentum F is no longer a good quantum number. However, m_F is still a good quantum number for it is associated with the conserved angular momentum of rotation about the magnetic field axis. It is then often convenient to label the single atom state with F and m_F, even in the high magnetic field limit, bearing in mind that F is the total angular momentum quantum number of the state at zero magnetic field, with which the present state adiabatically correlates.

the bound state as expansions of different channels

$$|\Psi(r, E)\rangle = |\Psi_\alpha\rangle + \sum_{\beta \neq \alpha} |\Psi_\beta(r, E)\rangle. \tag{6.29}$$

The Hamiltonian can be expressed in the matrix form

$$H = \begin{pmatrix} -\frac{\hbar^2 \nabla_r^2}{2\mu} + V_{\alpha\alpha}(r) & V_{\alpha\beta_1}(r) & V_{\alpha\beta_2}(r) \cdots \\ V_{\alpha\beta_1}^\dagger(r) & -\frac{\hbar^2 \nabla_r^2}{2\mu} + E_{\beta_1} + V_{\beta_1\beta_1}(r) & V_{\beta_1\beta_2}(r) \cdots \\ \vdots & \vdots & \vdots & \ddots \end{pmatrix}. \tag{6.30}$$

The diagonal elements of $V_{ii}(r)$ depend on weighted sums of the interacting potentials in the singlet $V_S(r)$ and triplet $V_T(r)$ channels, while the off-diagonal terms $V_{ij}(r)$ depend on the difference between $V_S(r)$ and $V_T(r)$. The energy shift of each channel E_{β_i} is in general a function of a magnetic field due to the Zeeman shift of different hyperfine states.

If the s-wave scattering process is allowed by symmetry, the low-energy scattering process is dominated by the s-wave contribution, which can be extracted by solving the Hamiltonian Eq. (6.30), together with the asymptotic boundary condition for the radial wavefunction

$$R_\alpha(r, E) = \frac{1}{kr} \sin{(kr + \delta_0)}. \tag{6.31}$$

In general, the resulting phase shift δ_0, or equivalently the s-wave scattering length a_s, is only weakly dependent on the magnetic-field strength B, unless B can be tuned in such a way that a bound state of a closed channel β_c crosses the threshold of the open-channel α. This can be achieved if the magnetic moments

$$m_\alpha = \partial E_\alpha / \partial B, \quad m_{\beta_c} = \partial E_{\beta_c} / \partial B \tag{6.32}$$

associated with open and closed channels are different.

If this is the case, the Hamiltonian Eq. (6.30) can be approximated by a two-channel model, which involves only the open-channel and closed-channel supporting the nearly resonant bound state

$$H_{2c} = \begin{pmatrix} -\frac{\hbar^2 \nabla_r^2}{2\mu} + V_{\alpha\alpha}(r) & V_{\alpha\beta_c}(r) \\ V_{\alpha\beta_c}^\dagger(r) & -\frac{\hbar^2 \nabla_r^2}{2\mu} + \nu_0 + \Delta m(B - B_0) + V_{\beta_c\beta_c}(r) \end{pmatrix}. \tag{6.33}$$

Here, $\Delta m \equiv m_{\beta_c} - m_\alpha$ is defined as the magnetic moment difference and ν_0 is the energy associated with the closed-channel at magnetic field B_0, at which the position of the closed-channel bound state is degenerate with the open-channel threshold. This model is qualitatively illustrated in Fig. 6.5. The magnetic field strength B thus plays the role of energy shift ν as in Eq. (6.10). As a result, the scattering length becomes a function of B

$$a_s(B) = a_{\text{bg}} \left(1 - \frac{\Delta B}{B - B_0} \right) \qquad (6.34)$$

with ΔB the resonance width and the bound state energy still takes the form as in Eq. (6.27).

The first wave of experiments demonstrating that the magnetically induced Feshbach resonance could be used to tune the interatomic interaction appeared in 1998 [7, 8]. Using a BEC of ^{23}Na, Inouye et $al.$ highlighted the two most striking features of a Feshbach resonance: the tunability of the scattering length according to Eq. (6.34) and the fast atom loss in the resonance region (see Fig. 6.6) [7]. The latter is then attributed to a strongly enhanced three-body recombination rate and molecule formation in the vicinity of a Feshbach resonance. In the same year, Courteille et $al.$ reported a Feshbach resonance in a magnetically trapped thermal sample of ^{85}Rb, by observing an enhanced photoassociative loss [8]. This result confirmed

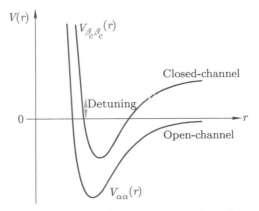

Figure 6.5.   An illustration of the two-channel model Eq. (6.33) for a magnetic Feshbach resonance. The detuning between the open-channel threshold and the closed-channel bound state energy can be tuned magnetically if the magnetic moments of the open and closed channels are different.

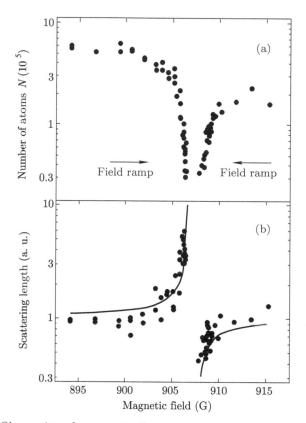

Figure 6.6. Observation of a magnetically tuned Feshbach resonance in an optically trapped BEC of ^{23}Na atoms. The upper panel depicts a severe loss of atoms near the resonance, which is due to the strongly enhanced three-body recombination rate. The lower panel shows the variation of scattering length by sweeping the magnetic field near resonance, as determined from measurements of the mean-field interaction energy by expansion of the condensate after it is released from the trap. Here, a_s is renormalized to the background value a_{bg}. From Ref. [7].

the prediction of using photoassociation as a probe to identify Feshbach resonances [42, 43]. The first wave of experiments immediately initiated a substantial wave of research in this field [41, 44]. At present, magnetic Feshbach resonances have been experimentally observed in essentially all alkali-metal species, in some mixtures of different alkali-metal atoms, as well as in Cr atoms. The Feshbach resonance has become a fully established tool and opened up many new applications in the field of cold atom physics.

6.3.   Optical Feshbach resonance

In addition to the magnetically tunable Feshbach resonance, where the energy difference between the closed-channel bound state and the open-channel threshold is tuned by an external magnetic field via the Zeeman effect, a resonance phenomenon can also be realized by applying a laser light, which induces resonance coupling between the two channels. A schematic illustration of such a scenario is shown in Fig. 6.7. For many species of atoms, the laser needed to produce an effective coupling is in the frequency domain of visible light. Thus, the term *optical Feshbach resonance* is widely used since it was first proposed by Fedichev *et al.* [45].

In this scheme, the open channel is constructed by two colliding atoms, both residing in the electronic ground state $|g\rangle$. The closed channel consists of one atom in the ground state and the other in an electronic excited level $|e\rangle$. In the closed channel, a rovibrational molecular state is selected as the bound state. By applying a laser light at frequency ω_ℓ, which is nearly resonant with the transition $\hbar\omega_c$ from the open-channel threshold to the molecular state in the closed channel, the two channels are effectively coupled. In the dressed-state picture as discussed in Sec. 3.3.2, the energy difference that is about to be tuned through resonance is $\Delta E = \hbar(\omega_\ell - \omega_c)$. Hence,

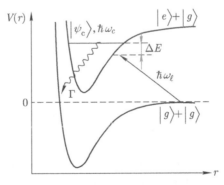

Figure 6.7.   A schematic diagram of an optical Feshbach resonance. Two colliding ground state atoms $|g\rangle$ are coupled to a rovibrational molecular level $|\psi_c\rangle$ of an electronically excited state atom $|e\rangle$ and a ground state atom $|g\rangle$ via a laser at frequency ω_ℓ. The energy of the rovibrational state relative to the threshold of the two colliding ground state atoms is denoted as $\hbar\omega_c$. The excited molecular state $|\psi_c\rangle$ is metastable only and is subjected to spontaneous decay at rate Γ. In the dressed state picture, the optical detuning between the open-channel threshold and the closed-channel bound state is $\Delta E = \hbar(\omega_\ell - \omega_c)$, which can be tuned by the laser frequency.

one can tune through an optical Feshbach resonance by sweeping the laser frequency ω_ℓ.

There is, however, a crucial difference between magnetic and optical Feshbach resonances. The excited molecular level employed in an optical Feshbach resonance is only a metastable state with a finite decay rate Γ due to spontaneous emission. Thus, the dispersion energy of this bound state acquires an imaginary part, which in turn leads to a complex scattering length in the open channel [45, 46]

$$\tilde{a}_s = a_{\rm bg}\left[1 + \frac{W_0(I)}{\hbar(\omega_\ell - \omega_c) + i\Gamma/2}\right]. \tag{6.35}$$

Here, the resonance width $W_0(I)$ is proportional to the inter-channel coupling strength. Unlike the case of a magnetic Feshbach resonance where the inter-channel coupling is fixed by interatomic interaction potentials, the optical case $W_0(I)$ can be turned on and off by changing the laser intensity I. In the vicinity of the resonance, the real and imaginary parts of the scattering length take the following forms

$$\text{Re}[\tilde{a}_s] = a_{\rm bg} + a_{\rm res}\frac{\Gamma\Delta E}{\Delta E^2 + \Gamma^2/4}, \tag{6.36a}$$

$$\text{Im}[\tilde{a}_s] = \frac{a_{\rm res}}{2}\frac{\Gamma^2}{\Delta E^2 + \Gamma^2/4}, \tag{6.36b}$$

where the resonant length parameter is defined as

$$a_{\rm res} \equiv a_{\rm bg}W_0/\Gamma. \tag{6.37}$$

One may immediately see from Eq. (6.36a) that the s-wave scattering length does not diverge even at the resonance point. Instead, the maximum value of $|\text{Re}[\tilde{a}_s]|$ is governed by the finite spontaneous decay rate Γ, as shown in Fig. 6.8.

In 2000, the optical Feshbach resonance was demonstrated in a system of trapped cold atom gas of sodium [47]. In their experiment, Fatemi et al. observed the changing scattering length by measuring the corresponding change in the scattering wavefunction. They found an $a_{\rm res}$ of around 2 nm and $W_0/\hbar = 2\pi \times 20\,\text{MHz}$ for the maximum laser intensity of $I = 100$ W/cm^2. In 2004, Theis et al. [48] tuned an optical Feshbach resonance in a ^{87}Rb BEC to achieve a variation of scattering length from 0.5 to 10 nm. In the next year, Thalhammer et al. [49] also implemented the resonance in rubidium BEC via a two-beam Raman transition.

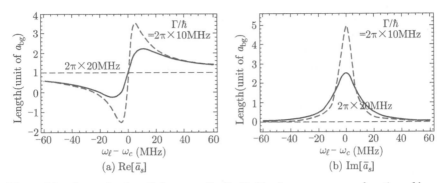

Figure 6.8.   Scattering length for an optically Feshbach resonance as a function of laser detuning $\omega_\ell - \omega_c$. The real and imaginary parts of a_s are defined as in Eq. (6.36). Parameters used here are $a_{\rm bg} = a_0$ and $W_0/\hbar = 2\pi \times 25\,{\rm MHz}$.

In order for an optical Feshbach resonance to be practical, it is necessary that the variation of the real part of scattering length $\mathrm{Re}[\tilde{a}_s] - a_{\rm bg}$ near resonance is much larger than the magnitude of the imaginary part $\mathrm{Im}[\tilde{a}_s]$, which defines the collisional loss rate. From the expressions of Eq. (6.36), this translates to the condition $a_{\rm res} \gg a_{\rm bg}$, or equivalently $W_0 \gg \Gamma$. This condition does not hold in existing experiments using alkali-metal atoms. As a consequence, systems suffer severe atom loss by tuning toward the optical Feshbach resonance [47–49].

One way to overcome this issue is to choose alkaline-earth atoms, instead of the commonly used alkali-metal systems, as proposed by Ciurylo et al. [50]. The crucial feature of alkaline-earth systems is the presence of intercombination lines, which are transitions between the ground $^1S_0$ state and the excited $^3P_1$ state. The transition is only weakly allowed due to its relativistic mixing with the $^1P_1$ state. For example, the linewidth of this transition for Sr isotopes is only $\Gamma/\hbar = 2\pi \times 7.5$ kHz, in contrast to $\Gamma/\hbar \sim$ MHz in alkali-metal atoms. When Γ is small, it is possible to use excited bound states that are very close to the closed-channel dissociation threshold, while simultaneously maintaining the large detunings to suppress losses. As a consequence, the ratio W_0/Γ, as well as the ratio $a_{\rm res}/a_{\rm bg}$, can both be very large.

The first observation of an optical Feshbach resonance in alkaline-earth systems was obtained by Zelevinsky et al. using ^{88}Sr atoms [51]. They extracted photoassociation spectra near the intercombination line and found that the shallowest bound state of the excited channel has a resonance length $a_{\rm res} = 24\ \mu$m at $I = 1$ W/cm^2. Similar photoassociation spectra have also been observed for two different isotopes of ytterbium [52], which has an electronic structure like that of alkaline-earth atoms. In 2008, Enomoto *et al.* [53] demonstrated that optical Feshbach resonances can be used to tune scattering length for ^{172}Yb and ^{176}Yb. More recently, the interaction between ^{88}Sr atoms is tuned by optical means to control the collapse and expansion of the condensate [54].

References

[1] U. Fano. *Phys. Rev.* **124**, 1866 (1961).
[2] H. Feshbach. *Ann. Phys. (N. Y.)* **5**, 357 (1958).
[3] H. Feshbach. *Ann. Phys. (N. Y.)* **19**, 287 (1962).
[4] W. C. Stalley. *Phys. Rev. Lett.* **37**, 1628 (1976).
[5] M. W. Reynolds, I. Shinkoda, R. W. Cline and W. N. Hardy. *Phys. Rev. B* **34**, 4912 (1986).
[6] E. Tiesinga, B. J. Verhaar and H. T. C. Stoof. *Phys. Rev. A* **47**, 4114 (1993).
[7] S. Inouye, M. R. Andrews, J. Stenger, H.-J. Miesner, D. M. Stamper-Kurn and W. Ketterle. *Nature* **392**, 151 (1998).
[8] P. Courteille, R. S. Freeland, D. J. Heinzen, F. A. van Abeelen and B. J. Verhaar. *Phys. Rev. Lett.* **81**, 69 (1998).
[9] L. Khaykovich, F. Schreck, G. Ferrari, T. Bourdel, J. Cubizolles, L. D. Carr, Y. Castin and C. Salomon. *Science* **296**, 1290 (2002).
[10] K. E. Strecker, G. B. Partridge, A. G. Truscott and R. G. Hulet. *Nature* **417**, 150 (2002).
[11] S. L. Cornish, N. R. Claussen, J. L. Roberts, E. A. Cornell and C. E. Wieman. *Phys. Rev. Lett.* **85**, 1795 (2000).
[12] T. Weber, J. Herbig, M. Mark, H.-C. Nägerl and R. Grimm. *Science* **299**, 232 (2003).
[13] T. Kraemer, J. Herbig, M. Mark, T. Weber, C. Chin, H.-C. Nägerl and R. Grimm. *Appl. Phys. B: Lasers Opt.* **79**, 1013 (2004).
[14] G. Roati, M. Zaccanti, C. D'Errico, J. Catani, M. Modugno, A. Simoni, M. Inguscio and G. Modugno. *Phys. Rev. Lett.* **99**, 010403 (2007).
[15] T. Loftus, C. A. Regal, C. Ticknor, J. L. Bohn and D. S. Jin. *Phys. Rev. Lett.* **88**, 173201 (2002).
[16] K. O'Hara, S. L. Hemmer, M. E. Gehm, S. R. Granade and J. E. Thomas. *Science* **298**, 2179 (2002).
[17] S. Jochim, M. Bartenstein, A. Altmeyer, G. Hendl, S. Riedl, C. Chin, J. Hecker Denschlag and R. Grimm. *Science* **302**, 2102 (2003).
[18] M. W. Zwierlein, C. A. Stan, C. H. Schunck, S. M. F. Raupach, S. Gupta, Z. Hadzibabic and W. Ketterle. *Phys. Rev. Lett.* **91**, 250401 (2003).
[19] T. Bourdel, L. Khaykovich, J. Cubizolles, J. Zhang, F. Chevy, M. Teichmann, L. Tarruell, S. J. J. M. F. Kokkelmans and C. Salomon. *Phys. Rev. Lett.* **93**, 050401 (2004).

[20] M. Greriner, C. A. Regal and D. S. Jin. *Nature* **426**, 537 (2003).

[21] C. A. Regal, M. Greiner and D. S. Jin. *Phys. Rev. Lett.* **92**, 040403 (2004).

[22] M. Bartenstein, A. Altmeyer, S. Riedl, S. Jochim, C. Chin, J. Hecker Denschlag and R. Grimm. *Phys. Rev. Lett.* **92**, 120401 (2004).

[23] J. Kinast, S. L. Hemmer, M. E. Gehm, A. Turlapov and J. E. Thomas. *Phys. Rev. Lett.* **92**, 150402 (2004).

[24] C. Chin, M. Bartenstein, A. Altmeyer, S. Riedl, S. Jochim, J. Hecker Denschlag and R. Grimm. *Science* **305**, 1128 (2004).

[25] G. B. Partridge, K. E. Strecker, R. I. Kamar, M. W. Jack and R. G. Hulet. *Phys. Rev. Lett.* **95**, 020404 (2005).

[26] M. W. Zwierlein, J. R. Abo-Shaeer, A. Schirotzek, C. H. Schunck and W. Ketterle. *Nature* **435**, 1047 (2005).

[27] A. J. Leggett. Diatomic molecules and Cooper pairs, in *Lecture Notes in Physics*. Eds. A. Pekalski and J. A. Przystawa. Spring Verlag, Berlin (1980).

[28] C. A. Regal, C. Ticknor, J. L Bohn and D. S. Jin. *Phys. Rev. Lett.* **90**, 053201 (2003).

[29] J. Zhang, E. G. M. van Kempen, T. Bourdel, L. Khaykovich, J. Cubizolles, F. Chevy, M. Teichmann, L. Tarruell, S. J. J. M. F. Kokkelmans and C. Salomon. *Phys. Rev. A* **70**, 030702(R) (2004).

[30] C. H. Schunck, M. W. Zwierlein, C. A. Stan, S. M. F. Raupach, W. Ketterle, A. Simoni, E. Tiesinga, C. J. Williams and P. S. Julienne. *Phys. Rev. A* **71**, 045601 (2005).

[31] M. Houbiers, H. T. C. Stoof, W. I. McAlexander and R. G. Hulet. *Phys. Rev. A* **57**, R1497 (1998).

[32] P. J. Leo, C. J. Williams and P. S. Julienne. *Phys. Rev. Lett.* **85**, 2721 (2000).

[33] A. Marte, T. Volz, J. Schuster, S. Dürr, G. Rempe, E. G. M. van Kempen and B. J. Verhaar. *Phys. Rev. Lett.* **89**, 283202 (2002).

[34] C. Chin, V. Vulentić, A. J. Kerman, S. Chu, E. Tiesinga, P. J. Leo and C. J. Williams. *Phys. Rev. A* **70**, 032701 (2004).

[35] B. Marcelis, E. G. M. van Kempen, B. J. Verhaar and S. J. J. M. F. Kokkelmans. *Phys. Rev. A* **70**, 012701 (2004).

[36] M. Bartenstein, A. Altmeyer, S. Riedl, R. Geursen, S. Jochim, C. Chin, J. Hecker Denschlag, R. Grimm, A. Simoni, E. Tiesinga, C. J. Williams and P. S. Julienne. *Phys. Rev. Lett.* **94**, 103201 (2005).

[37] M. S. Child. *Molecular Collision Theory.* Academic, London (1974).

[38] A. J. Moerdijk, B. J. Verhaar and A. Aexlsson. *Phys. Rev. A* **51**, 4852 (1995).

[39] E. Timmermans, P. Tommasini, M. Hussein and A. Kerman. *Phys. Rep.* **315**, 199 (1999).

[40] F. H. Mies and M. Raoult. *Phys. Rev. A* **62**, 012708 (2000).

[41] C. Chin, R. Grimm, P. Julienne and E. Tiesinga. *Rev. Mod. Phys.* **82**, 1225 (2010).

[42] J. M. Vogels, C. C. Tsai, R. S. Freeland, S. J. J. M. F. Kokkelmans, B. J. Verhaar and D. J. Heinzen. *Phys. Rev. A* **56**, R1067 (1997).

[43] F. A. van Abeelen, D. J. Heinzen and B. J. Verhaar. *Phys. Rev. A* **57**, R4102 (1998).

[44] T. Köhler, K. Córal and P. S. Julienne. *Rev. Mod. Phys.* **78**, 1311 (2006).
[45] P. O. Fedichev, Yu. Kagan, G. V. Shlyapnikov and J. T. M. Walraven. *Phys. Rev. Lett.* **77**, 2913 (1996).
[46] J. L. Bohn and P. S. Julienne. *Phys. Rev. A* **56**, 1486 (1997).
[47] F. K. Fatemi, K. M. Jones and P. D. Lett. *Phys. Rev. Lett.* **85**, 4462 (2000).
[48] M. Theis, G. Thalhammer, K. Winkler, M. Hellwig, G. Ruff, R. Grimm and J. Hecker Denschlag. *Phys. Rev. Lett.* **93**, 123001 (2004).
[49] G. Thalhammer, M. Theis, K. Winkler, R. Grimm and J. Hecker Denschlag. *Phys. Rev. A* **71**, 033403 (2005).
[50] R. Ciurylo, E. Tiesinga and P. S. Julienne. *Phys. Rev. A* **71**, 030701(R) (2005).
[51] T. Zelevinsky, M. M. Boyd, A. D. Ludlow, T. Ido, J. Ye, R. Ciurylo, P. Naidon and P. S. Julienne. *Phys. Rev. Lett.* **96**, 203201 (2006).
[52] S. Tojo, M. Kitagawa, K. Enomoto, Y. Kato, Y. Takasu, M. Kumakura and Y. Takahashi. *Phys. Rev. Lett.* **96**, 153201 (2006).
[53] K. Enomoto, K. Kasa, M. Kitagawa and Y. Takahashi. *Phys. Rev. Lett.* **101**, 203201 (2008).
[54] M. Yan, B. J. DeSalvo, B. Ramachandhran, H. Pu and T. C. Killian. *Phys. Rev. Lett.* **110**, 123201 (2013).

Part II

Ultracold Fermi Gases

7
Background and Experimental Achievements

Following the experimental realization of BEC, a natural step forward is to bring fermionic atoms into quantum degeneracy. For neutral noninteracting bosonic atoms, the quantum degeneracy condition $n\lambda_d^3 \sim 1$ leads to matter-wave interference and hence condensation. In contrast, such a phase transition does not exist for neutral noninteracting fermionic atoms, due to Fermi–Dirac statistics. This can be understood in the zero-temperature limit of noninteracting fermions, where atoms occupy all quantum states below a Fermi energy due to the Pauli blocking. This leads to a sharp Fermi surface in momentum space. It is well-known that the Fermi surface becomes unstable with the introduction of even an infinitesimally small attractive interaction between different atom species. Due to the so-called Cooper instability, the many-body ground state of a two-component Fermi gas with attractive interactions is not a simple Fermi sea, but rather a superfluid of correlated atom pairs, which are the analog of Cooper pairs of electrons in the BCS superconducting materials. On the other hand, the situation is more subtle for a Fermi gas with repulsive interactions. For example, when the repulsive interaction originates from attractive two-body scattering potentials, there exist different branches of many-body states that may lead to either a superfluid of two-body bound states or an effectively repulsive Fermi gas.

In this chapter, we will first review the experimental developments in the past decade with ultracold Fermi gases. These experiments reveal the rich physics within the underlying systems, which have important implications for a wide range of topics: high-T_c superconductors, magnetized superfluidity and quark matter, to name a few. We will then focus on Fermi gases with attractive interactions and introduce the general physical picture of a BCS-BEC crossover, which will be discussed in detail in the following chapters.

7.1. Brief introduction to experimental achievements

The unique behavior of quantum degenerate Fermi gases originates from the Pauli exclusion principle, which states that no two fermions can occupy the same quantum state. Therefore, in the zero-temperature limit, instead of condensing into the quantum state with the lowest energy, fermions occupy all available quantum states below the Fermi surface that is determined by the total density of fermions. This gives rise to many important effects, e.g., the stability of the electron shell structure in an atom, the Fermi pressure that keeps a neutron star from collapsing, etc. It is therefore intriguing that the properties of a degenerate Fermi gas can be studied in the context of ultracold fermionic atoms with highly tunable parameters. Furthermore, the successful generation of degenerate Fermi gases opens up the possibility of preparing and studying pairing superfluidity in cold atomic gases.

Compared to a Bose gas, a Fermi gas is more difficult to bring into quantum degeneracy. Due to the Pauli blocking, s-wave scattering between identical fermionic atoms is forbidden. A single-component Fermi gas is therefore difficult to thermalize, which makes evaporative cooling ineffective. A practical route to circumvent this difficulty is to introduce a second atom species which has a considerable s-wave scattering rate with the original component of the Fermi gas. In this so-called sympathetic cooling scheme, the same fermionic atoms in different hyperfine states, or even different bosonic atoms, are typically introduced, so that the Fermi gas may thermalize properly.

In 1999, D. S. Jin's group was the first to achieve the quantum degeneracy of a Fermi gas using ^{40}K atoms [1]. Using the $|F = 9/2, m_F = 9/2\rangle$ and $|F = 9/2, m_F = 7/2\rangle$ hyperfine states as the two different (pseudo-) spin species and performing sympathetic cooling, they were able to cool the gas to $\sim 0.5T_F$, where the Fermi temperature T_F is on the order of μK. Though this temperature is still higher than the normal-superfluid phase-transition temperature, the degenerate Fermi gas already behaves qualitatively differently from a classical gas. For example, the total energy for a classical gas consisting of N particles at temperature T is $3Nk_BT$. For a degenerate noninteracting Fermi gas, the total energy and the chemical potential at temperature T can be calculated from the Fermi–Dirac statistics

$$E = \int d\epsilon g(\epsilon) \frac{\epsilon}{e^{\beta(\epsilon-\mu)} + 1}, \qquad (7.1)$$

$$N = \int d\epsilon g(\epsilon) \frac{1}{e^{\beta(\epsilon-\mu)} + 1}, \qquad (7.2)$$

where $g(\epsilon)$ is the density of state and $\beta = 1/k_BT$. For fermions in a harmonic trap and at zero temperature, the chemical potential μ is determined by Eq. (7.2), which gives $E = 3Nk_BT_F/4$ when substituted into Eq. (7.1). Thus, by measuring the total energy of the degenerate gas via a time-of-flight (TOF) imaging process, one may see signatures of the quantum effects. Similar effects were also observed in ^{6}Li atoms later in 2001 at Ecole Normale Superieure (ENS) and at the Rice University, where sympathetic cooling was performed using atoms of ^{6}Li and their bosonic isotope ^{7}Li [2, 3]. In all these experiments, the final temperatures of the gas were in the quantum degenerate region, but still higher than the critical temperature for pairing superfluidity. One way to achieve that goal is to bring the gas into the strongly interacting regime, where the superfluid transition temperature is enhanced significantly. This can be realized for instance by tuning the two-body scattering length via the Feshbach resonance technique.

By that time, the ability to tune a two-body scattering length with the Feshbach resonance had already been studied in a series of experiments with ultracold Bose gases. In a cold gas with bosonic atoms, however, a large scattering length leads to strong three-body decay processes, during which three atoms collide to form a deeply bound dimer, while a free atom carries away the large binding energy [4, 5]. For example, in a 1999 experiment, a research group from the Massachusetts Institute of Technology (MIT) observed a loss of $\sim 70\%$ of atoms within $1\mu s$ while ramping the magnetic field across the 907G Feshbach resonance point of ^{23}Na [6]. This large three-body loss rate has been considered to prevent a Bose gas with strong interaction from being stabilized. Interestingly, a recent experiment has suggested that a strongly interacting Bose gas at unitarity can be prepared and probed on a timescale longer than its thermolization time, but shorter than its lifetime [7]. For an ultracold Fermi gas, however, this is not the case. The effects of the Pauli blocking inhibit three-body processes and thus lead to a much more stable Fermi gas in the strongly interacting regime. Theoretically, the three-body losses in ultracold atomic gases have been well characterized by Petrov *et al.* [8].

With the Feshbach resonance technique, the *s*-wave scattering length between different spin species of a Fermi gas can be tuned from being positive to negative, and from small to very large values. For small and negative scattering lengths, the ground state of the system is a weakly interacting Fermi gas which supports BCS-type pairing superfluidity at low enough temperatures. For small and positive scattering lengths, meta-stable bound states composed of two different fermionic atoms form and change the quantum statistics of the gas from being fermionic to bosonic. At low enough temperatures, these molecules may condense and form a BEC. This provides the opportunity to continuously tune the system from a BCS-type pairing superfluid to a molecular BEC. Between these two limiting cases, we may have a very large scattering length that diverges at the Feshbach resonance point. Near the resonance point, the scattering length becomes much larger than the inter-particle separation and we will have a unique system that is dilute and strongly correlated at the same time.

The first report of strongly interacting Fermi gas was by a research group from Duke University in 2002, where this group observed anisotropic expansion of the resonantly interacting Fermi gas after switching off the trapping potential [9]. This hydrodynamic behavior is in contrast to the isotropic collisionless expansions typical of a weakly interacting Fermi gas. Similar behavior was also reported by a research group from the ENS, where the interaction energy of the strongly interacting Fermi gas was also measured by comparing the TOF image of a strongly interacting gas and that of a non-interacting gas [10]. As the hydrodynamic expansion may imply either classical collisional hydrodynamics or superfluid hydrodynamics (as in a BEC), it is not an unambiguous signal of the superfluid phase. In fact, the anisotropic expansion in both experiments were interpreted as mere collisional hydrodynamics.

The creation of bosonic molecules on the BEC side of the Feshbach resonance was first reported in early 2003 by a research group from the JILA, where this group prepared a degenerate Fermi gas of ^{40}K and adiabatically tuned the magnetic field across the Feshbach resonance to form molecules composed of $|F = 9/2, m_F = -9/2\rangle$ and $|F = 9/2, m_F = -7/2\rangle$ hyperfine spin states [11]. This was followed by separate research groups from

Rice University, ENS and Innsbruck University, where molecules were formed using ^{6}Li atoms [12, 13, 15]. Remarkably, while the measured lifetime of the composite bosons in ^{40}K was on the order of 1 ms, the lifetime in ^{6}Li was on the order of 10 s. Later on in the same year, research groups from JILA, Innsbruck University and MIT reported the formation of Bose–Einstein condensation of molecules from ultracold Fermi gases, where bimodal density distributions typical of a BEC had been observed [16–19]. At JILA, for ^{40}K atoms, the molecules and hence the molecular BEC were prepared by evaporative cooling on the BCS side of the Feshbach resonance, followed by an adiabatic sweeping of the magnetic field to the BEC side. For ^{6}Li, the strategy was to directly cool the gas on the BEC side (see Fig. 7.1). As the two-body process leading to molecular formation is exoergic, the molecules are favored at low temperatures. These composite bosons will further condense to form the BEC once the temperature becomes low enough. The reason behind these different strategies lies in the fact that ^{6}Li molecules are more stable in the BEC region, and in the case of ^{6}Li the BEC side of the resonance can offer significantly more quantum

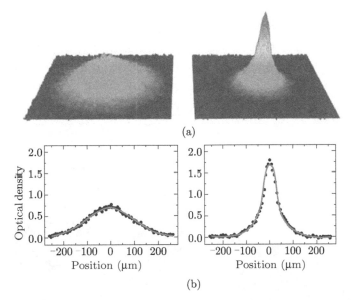

Figure 7.1. Molecular BEC emerging from a Fermi sea. As the temperature is lowered, a bimodal structure shows up in the momentum distribution of the molecules on the BEC side of the resonance. From Ref. [16].

states for a fixed trapping potential, which leads to many more atoms in the trap [14].

Soon after the generation of molecular BEC, the JILA and MIT research groups reported pairing superfluidity in the strongly interacting Fermi gas on the BCS side of the resonance [20, 21]. In these seminal experiments, the pairing correlations as well as the momentum distribution of the fermion pairs in a strongly interacting Fermi gas were projected onto the molecular states on the BEC side by a sudden quench of the magnetic field (see Fig. 7.2). The resulting bimodal structure in the molecular density distribution demonstrated the superfluid nature of the fermionic atom pairs in the initial state. These experiments stimulated tremendous excitement in the fields of ultracold atoms and condensed matter physics. Besides generating novel strongly correlated quantum many-body states, they provided the opportunity to study in detail the crossover from BEC to BCS superfluidity, especially in the strongly interacting region.

During the following years various groups systematically studied the properties of the BCS-BEC crossover in ultracold Fermi gases. These included measurements of the release energy [19], the density profiles [22], the thermodynamic properties [23], the population in the closed channel, and so on and so forth [24]. People were looking to confirm the superfluid nature of the gas at low temperatures in the crossover region, as well as to further understand the properties of the system. In particular, collective excitation of the Fermi gas throughout the crossover region was measured

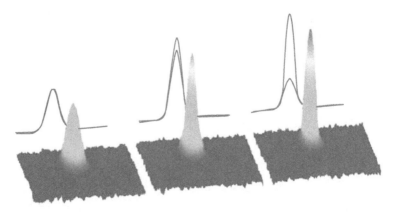

Figure 7.2.   Projection of the fermion pair condensate onto the molecular BEC. From Ref. [21].

by the Duke and Innsbruck research groups, where a breakdown of the hydrodynamic behavior toward the BCS side suggests a superfluid-normal phase transition. Another seminal experiment was the measurement of the pairing gap via radio-frequency spectroscopy (see Fig. 7.3). By coupling the hyperfine state of one of the spin species to a bystander hyperfine state via radio-frequency pulses and then measuring the resonant frequency, the Innsbruck research group was able to extract information regarding the binding energy of the atom pairs [25]. Although the background scattering between the other spin species and the bystander state rendered the experimental data difficult to analyze quantitatively, they still provided important clues to the pairing correlations throughout the crossover region and at different temperatures [26]. The final piece of evidence for the resonant superfluidity was provided by the MIT research group in 2005, when they created and observed quantized vortices throughout the crossover region [27] (see Fig. 7.4).

In 2006, experiments at MIT and Rice University on the quantum phase transition in a spin-polarized Fermi gas near a wide Feshbach resonance attracted much attention (see Fig. 7.5) [28, 29]. In contrast to the spin balanced case, where tuning across the Feshbach resonance leads to a smooth crossover from BCS superfluidity to BEC, a quantum phase transition occurs due to the competition between pairing and the population imbalance of the two spin species. Furthermore, exotic phases like the

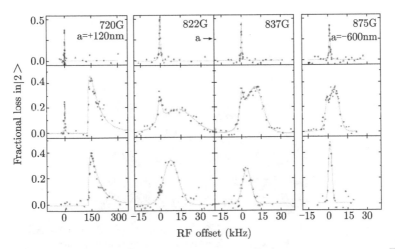

Figure 7.3.   Measurement of the pairing gap using radio-frequency spectroscopy. From Ref. [25].

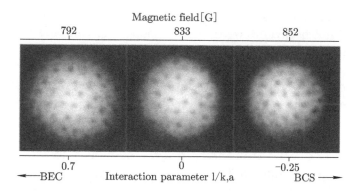

Figure 7.4.   Formation of vortices throughout the BCS-BEC crossover. From Ref. [27].

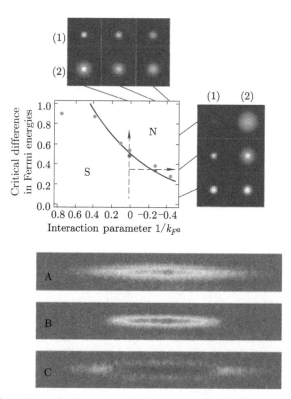

Figure 7.5.   Experimental observation of phase transition and phase separation in a polarized Fermi gas. (a) Phase transition and density distribution in the MIT experiment. From Ref. [28]. (b) Density distribution from the Rice experiment. From Ref. [29].

breached pairing phase and the Fulde–Ferrell–Larkin–Ovchinnikov (FFLO) phase may be stabilized as a result of population imbalance [30]. These again stimulated intense theoretical and experimental investigations, which provided interesting insights into related systems in the contexts of other research fields, e.g., magnetized superfluid, quark matter, neutron star, etc.

Another important direction is the study of a Fermi gas in a periodic optical lattice potential. As has been suggested in the seminal paper by Jaksch *et al.* [31], such configurations have the potential to simulate important models in condensed matter physics, e.g., the Bose–Hubbard model, the Fermi–Hubbard model, etc. Experimentalists are now working hard to bring the lattice fermions into temperatures low enough for the novel phases and interesting physics to show up. We will discuss these in Part III. For now, we will focus on attractively interacting Fermi gases in a non-periodic external trapping potential.

7.2.   BCS-BEC crossover

The BCS-BEC crossover problem was first considered by Eagles in the 1960s [32] and later by Leggett in the 1980s, [33] when they tried to connect the physical origins of superfluidity and superconductivity. It was suggested that the BCS wave function, which describes the ground state of a fermionic system in the weak-coupling limit, could be applied to fermionic systems with arbitrary coupling strengths. Therefore, the long-range Cooper pairs, which are responsible for the superconductivity in the weak-coupling limit, can also be regarded as spatially overlapping composite bosons, which condense below the superconducting temperature. In the strong-coupling limit, however, the fermions form tightly bound pairs and the resultant composite bosons condense into a molecular BEC below the condensation temperature. The BCS-BEC crossover theory predicts a smooth crossover from the BCS ground state in the weak-coupling limit to the BEC ground state in the strong-coupling limit, as the coupling strength is adiabatically varied from one limit to the other. The general physical picture of the BCS-BEC crossover is illustrated in Fig. 7.6. In the crossover region, fermion pairs are intermediate in range and strongly interact with each other, such that the properties of the system are quite different from those in the BCS or in the BEC limit. While these two limiting cases can be well described using a mean-field theory, the intermediate crossover region proves to be difficult to characterize [34].

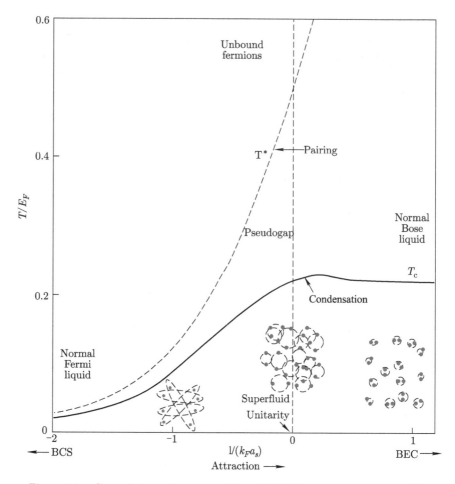

Figure 7.6.   General physical picture of the BCS-BEC crossover. From Ref. [35].

From the physical picture above, it is easy to identify the BEC side of the Feshbach resonance as the strong-coupling limit and the BCS side as the weak-coupling limit. Indeed, we see that with a positive s-wave scattering length, i.e., on the BEC side of the resonance, the two-body interaction supports a bound dimer state. Although this molecular dimer is not the ground state of the system, its binding energy is still much larger than other relevant energy scales of the system. Therefore, the size of these molecules is much smaller than the inter-particle separation. The many-body problem in this limit can therefore be considered as weakly interacting bosonic molecules. On the BCS side of the resonance, according to the BCS theory, atoms with opposite spins form Cooper pairs near the Fermi surface. As the "binding energy" of the Cooper pairs is much smaller than the Fermi energy, the size of these Cooper pairs is much larger than the inter-particle separation. This leads to a physical picture of weakly-interacting (non-interacting in the mean-field description) Cooper pairs. One may then imagine a smooth crossover connecting these two limiting cases, where the typical size of the atom pairs changes smoothly. In the strongly interacting region, the typical size of the atom pairs is on the order of the inter-particle separation and the system is strongly correlated.

At the resonance point, the two-body scattering length diverges. The detailed two-body short-range interaction potential corresponds to a very large energy scale, which becomes irrelevant in the low-energy regime where most of the interesting physics takes place. Thus, for a uniform Fermi gas at zero temperature, the only relevant energy scale is the Fermi energy and the only relevant length scale is the inter-particle separation. At zero temperature, this implies that physical quantities like the chemical potential or the effective mass can be related to those of a non-interacting Fermi gas by a simple scaling relation. At finite temperatures, all the thermodynamical properties should be functions of E_F and T/T_F, regardless of the detailed two-body interaction potentials. This is the so-called unitary limit of the gas [36]. One of the most amazing results of this unitary behavior is that the transition temperature in the unitary region is of the same order as the Fermi temperature, with $T_c \sim 0.2T_F$ [37]. It is this significant enhancement of the transition temperature that allows for the experimental observation of the Fermi condensate. Besides, many theorists have associated this extremely high transition temperature with that of a high-T_c superconductor [34, 37]. Hence, the characterization of the BCS-BEC crossover in ultracold Fermi gases provides valuable insights into the long-standing problem of high-T_c superconductivity. A highly non-trivial example in this respect is the study of pseudogap in the strongly interacting region, which concerns the finite temperature behavior of the Cooper pairs. It has been suggested that above the superfluid transition temperature T_c, there exists another critical temperature at which the fermions start to form correlated pairs. These pairs of atoms will eventually condense into Cooper pairs at T_c and become a superfluid. In the BCS limit, the pseudogap temperature coincides with the phase-transition temperature, while the difference between these two temperatures becomes appreciable toward the unitarity and the BEC side. There have been various experimental indications for the existence of pseudogap at unitarity [25, 38]. However, a quantitatively accurate characterization of the pseudogap and of the critical temperature for the onset of pre-formed pairs are difficult to get. A schematic phase diagram including the pseudogap state is shown in Fig. 7.6.

As we may expect, the BCS framework, being a mean-field theory, should provide good description of a system with weak interactions. In the BEC limit, the fermions form tightly bound molecules, whereas the interaction between the molecules is rather weak. In this sense, the mean-field calculation should provide quantitatively good descriptions both in the strong-coupling BEC limit and in the weak-coupling BCS region.[1] In the intermediate region, i.e., the BCS-BEC crossover region, the quantum fluctuations become important and the BCS-BEC crossover theory formulated by Eagles and Leggett can only provide a qualitatively valid picture. The quantum fluctuation in the strongly interacting region is carried by fermion pairs with non-zero center-of-mass momenta, whose contribution has to be taken into account for a quantitative description.

Ever since the pioneering works of Eagles and Leggett, many schemes have been put forth in order to provide beyond-the-mean-field descriptions of the BCS-BEC crossover. In 1985, Nozieres, and Schmitt-Rink (NSR) developed a T-matrix scheme to calculate the transition temperature T_c by including the self-energy of the quasi-particles in the number equation ($G_0 G_0$ scheme) [39]. Physically, this scheme includes the quantum fluctuation by a Gaussian expansion near the critical point of the superfluid-normal phase transition. In 1997, Engelbrecht *et al.* extended this scheme to below T_c, by making the Gaussian expansion near the saddle point (the broken symmetry phase) [40]. Other many-body schemes include the $G_0 G$ scheme and the GG scheme, both of which are variations based on the T-matrix scheme [34, 41]. Besides these efforts, alternative many-body schemes have been proposed recently [42, 43] and calculations with the Quantum Monte Carlo algorithm have been performed [44, 45]. Although there are still some small quantitative differences between the results of the different many-body calculations, the qualitative physical picture provided by the

---

[1]In the strong-coupling limit, the mean-field calculation provides quantitatively good description of the system, so long as the s-wave scattering length between the molecules is set to $a_d = 0.6a_s$, the result from exact few-body calculations [8].

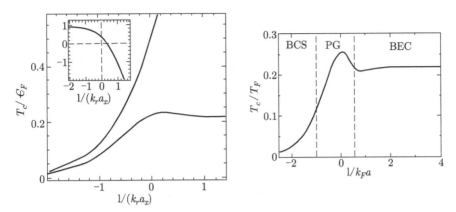

Figure 7.7. Critical temperature as a function of $1/(k_F a_s)$ from various theoretical calculations. (a) Results from NSR (solid line) versus results from a mean-field calculation (dashed line). From Ref. [40]. (b) Results from the $G_0 G$ scheme. From Ref. [34].

many-body calculations are consistent with that from a mean-field theory in most cases.

7.3. Overview

In the following chapters, we will discuss various aspects of the ultracold Fermi gases near a Feshbach resonance.

In Chap. 8, we first give a brief introduction of the BCS theory, which will serve as the foundation for the ensuing discussions on the BCS-BEC crossover. We keep our discussion on the BCS theory to a minimum here, while emphasizing the difference between a condensed matter system and an ultracold atomic gas. We then introduce the mean-field description of the BCS-BEC crossover theory based on the BCS wave function. Within this simple framework, we will discuss various properties of the fermionic atom pairs throughout the crossover region. To characterize properties of the strongly interacting Fermi gas in trapping potentials, we also discuss the local-density approximation (LDA) and the Bogoliubov–de-Gennes (BdG) equations in this chapter.

In Chap. 9, we introduce several schemes that are beyond the simple mean-field treatment. In the unitary regime, the system does not have any small parameters. Hence these schemes employ different strategies to include the fluctuations above the mean-field solution. Although there is no a priori reason to believe that they should produce accurate results, they typically fit well with experimental observations.

In Chaps. 10 and 11, we review two topics in ultracold Fermi gases that have attracted much attention in recent years: the polarized Fermi gas and the synthetic gauge field. We discuss the exotic phases and phase transitions in these systems and show that many of the interesting properties can be revealed on the simple mean-field level.

References

[1] B. De Marco and D. S. Jin. *Science* **285**, 1703 (1999).
[2] F. Schreck, L. Khaykovich, K. L. Corwin, G. Ferrari, T. Bourdel, J. Cubizolles and C. Salomon. *Phys. Rev. Lett.* **87**, 080403 (2001).
[3] A. K. Truscott, K. E. Strecker, W. I. McAlexander, G. B. Partridge and R. G. Hulet, *Science* **291**, 2570 (2001).
[4] P. Courteille, R. S. Freeland, D. J. Heinzen, F. A. van Abeelen and B. J. Verhaar. *Phys. Rev. Lett.* **81**, 69 (1998).
[5] S. Inouye, K. B. Davis, M. R. Andrews, J. Stenger, H. -J. Miesner, D. M. Stamper-Kurn and W. Ketterle. *Nature* **392**, 151 (1998).
[6] J. Stenger, S. Inouye, M. R. Andrews, H.-J. Miesner, D. M. Stamper-Kurn and W. Ketterle. *Phys. Rev. Lett.* **82**, 2422 (1999).
[7] P. Makotyn, C. E. Klauss, D. L. Goldberger, E. A. Cornell and D. S. Jin. *Nat. Phys.* **10**, 116 (2014).
[8] D. S. Petrov, C. Salomon and G. V. Shlyapnikov. *Phys. Rev. Lett.* **93**, 090404 (2004).
[9] K. M. O'Hara, S. L. Hemmer, M. E. Gehm, S. R. Granade and J. E. Thomas. *Science* **298**, 2179 (2002).
[10] T. Bourdel, J. Cubizolles, L. Khaykovich, K. M. F. Magalhães, S. J. J. M. F. Kokkelmans, G. V. Shlyapnikov and C. Salomon. *Phys. Rev. Lett.* **91**, 020402 (2003).
[11] C. A. Regal, C. Ticknor, J. L. Bohn and D. S. Jin. *Nature* **424**, 47 (2003).
[12] K. E. Strecker, G. B. Partridge and R. G. Hulet. *Phys. Rev. Lett.* **91**, 080406 (2003).
[13] J. Cubizolles, T. Bourdel, S. J. J. M. F. Kokkelmans, G. V. Shlyapnikov and C. Salomon. *Phys. Rev. Lett.* **91**, 240401 (2003).
[14] R. Grimm. *Proceedings of the International School of Physics 'Enrico Fermi'*, Course CLXIV. Eds. M. Inguscio, W. Ketterle and C. Salomon, Varenna, 20 to June 2006.
[15] S. Jochim, M. Bartenstein, A. Altmeyer, G. Hendl, C. Chin, J. Hecker Denschlag and R. Grimm. *Phys. Rev. Lett.* **91**, 240402 (2003).
[16] M. Greiner, C. A. Regal and D. S. Jin. *Nature* **426**, 537 (2003).
[17] S. Jochim, M. Bartenstein, A. Altmeyer, G. Hendl, C. Chin, J. H. Denschlag and R. Grimm. *Science* **302**, 2101 (2003).

[18] M. W. Zwierlein, C. A. Stan, C. H. Schunck, S. M. F. Raupach, S. Gupta, Z. Hadzibabic and W. Ketterle. *Phys. Rev. Lett.* **91**, 250401 (2003).

[19] T. Bourdel, L. Khaykovich, J. Cubizolles, J. Zhang, F. Chevy, M. Teichmann, L. Tarruell, S. J. J. M. F. Kokkelmans and C. Salomon. *Phys. Rev. Lett.* **93**, 050401 (2004).

[20] C. A. Regal, M. Greiner and D. S. Jin. *Phys. Rev. Lett.* **92**, 040403 (2004).

[21] M. W. Zwierlein, C. A. Stan, C. H. Schunck, S. M. F. Raupach, A. J. Kerman and W. Ketterle. *Phys. Rev. Lett.* **92**, 120403 (2004).

[22] M. Bartenstein, A. Altmeyer, S. Riedl, S. Jochim, C. Chin, J. Hecker Denschlag and R. Grimm. *Phys. Rev. Lett.* **92**, 120401 (2004).

[23] J. Kinast, A. Turlapov, J. E. Thomas, Q. Chen, J. Stajic and K. Levin. *Science* **307**, 1296 (2005).

[24] G. B. Partridge, K. E. Strecker, R. I. Kamar, M. W. Jack and R. G. Hulet. *Phys. Rev. Lett.* **95**, 020404 (2005).

[25] C. Chin, M. Bartenstein, A. Altmeyer, S. Riedl, S. Jochim, J. H. Denschlag and R. Grimm. *Science* **305**, 1128 (2004).

[26] C. H. Schunck, Y. Shin, A. Schirotzek and W. Ketterle. *Nature* **454**, 739 (2010).

[27] M. W. Zwierlein, J. R. Abo-Shaeer, A. Schirotzek, C. H. Schunck and W. Ketterle. *Nature* **435**, 1047 (2005).

[28] M. W. Zwierlein, A. Schirotzek, C. H. Schunck and W. Ketterle. *Science* **311**, 492 (2006).

[29] G. B. Patridge, W. Li, R. I. Kamar, Y. Liao and R. G. Hulet. *Science* **311**, 496 (2006).

[30] D. E Sheehy and L. Radzihovsky. *Ann. Phys.* **322**, 1790 (2007).

[31] D. Jaksch, C. Bruder, J. I. Cirac, C. W. Gardiner and P. Zoller. *Phys. Rev. Lett.* **82**, 1975 (1998).

[32] D. M. Eagles. *Phys. Rev.* **186**(2), 456 (1969).

[33] A. J. Leggett. *Modern Trends in the Theory of Condensed Matter.* Springer-Verlag, Berlin.

[34] Q. Chen, J. Stajic, S. Tan and K. Levin. *Phys. Rep.* **412**, 1 (2005).

[35] M. Randeria. *Nat. Phys.* **6**, 561 (2010).

[36] T.-L. Ho. *Phys. Rev. Lett.* **92**, 090402 (2004).

[37] M. Holland, S. J. J. M. F. Kokkelmans, M. L. Chiofalo and R. Walser. *Phys. Rev. Lett.* **87**, 120406 (2001).

[38] J. P. Gaebler, J. T. Stewart, T. E. Drake, D. S. Jin, A. Perali, P. Pieri and G. C. Strinati. *Nat. Phys.* **6**, 569 (2010).

[39] P. Nozières and S. Scmitt-Rink, J. Low Temp. *Phys.* **59**, 195 (1985).

[40] C. A. R. Sá de Melo, M. Randeria and J. R. Engelbrecht. *Phys. Rev. Lett.* **71**, 3202 (1993).

[41] R. Haussmann. Z. *Phys. B* **91**(3), 291, (1993).

[42] H. Hu, X.-J. Liu and P. D. Drummond. *Europhys. Lett.* **74**, 574 (2006).

[43] S. Diehl and C. Wetterich. *Phys. Rev. A* **73**, 033615 (2006).

[44] J. Carlson, S.-Y. Chang, V. R. Pandharipande and K. E. Schmidt. *Phys. Rev. Lett.* **91**, 050401 (2003).

[45] A. Bulgac, J. E. Drut and P. Magierski. *Phys. Rev. Lett.* **96**, 090404 (2006).

[46] S. Tan. *Ann. Phys.* **323**, 2971 (2008).

[47] S. Tan. *Ann. Phys.* **323**, 2987 (2008).

[48] J. T. Stewart, J. P. Gaebler, T. E. Drake and D. S. Jin. *Phys. Rev. Lett.* **104**, 235310 (2010).

[49] R. J. Wild, P. Makotyn, J. M. Pino, E. A. Cornell and D. S. Jin. *Phys. Rev. Lett.* **108**, 145305 (2012).

8
BCS-BEC Crossover

In this chapter, we introduce the mean-field theory of the BCS-BEC crossover as a natural extension of the standard BCS theory, while emphasizing their critical differences. Our introduction focuses on wide Feshbach resonances, which are studied by most of the experiments. For completeness, we also briefly discuss cases of narrow Feshbach resonance.

8.1. Cooper instability

For electrons in a metal, the effective interaction between electrons have two major contributions: the screened Coulomb interaction and the screened electron-phonon interaction [1]. While the screened Coulomb interaction is repulsive in the low-energy limit, which is the case at low-temperatures, the electron-phonon interaction can become large and attractive near the Fermi surface. A qualitative understanding of the effective interaction between electrons is the following: an electron moving in the presence of a lattice of ions causes vibrations (phonon modes) in the lattice, which in turn affects the motion of electrons with the opposite spin and momentum. This results in an effective, attractive interaction between the electrons. This interaction modifies the ground state of the metal in a fundamental way.

As we have discussed previously, for a non-interacting Fermi gas at zero temperature, the ground state is a simple Fermi sea, where all the quantum states below the Fermi surface are occupied. For electrons in a real metal, considering the effective attraction between them, this simple picture is drastically changed. In 1956, Cooper pointed out that the ground state of a normal metal would become unstable at zero temperature; the existence of small attractive interactions would destabilize the Fermi surface [2]. The electrons near the Fermi surface interact with those on the opposite side of the Fermi sea and develop bound states, thus rendering the original Fermi sea unstable. In the following, we will discuss a simple model that captures the physical picture here.

Let us consider two electrons with different spins above a fully occupied Fermi sea at zero temperature. Neglecting interaction between the two electrons and the electrons in the Fermi sea below, this is effectively a two-body problem in the presence of a Fermi sea, i.e., with modified density of states. As we have done before, we may separate the center-of-mass motion and consider only the relative motion of the two electrons. The Schrödinger's equation for the wave function of the relative motion is

$$-\frac{\hbar^2}{m}\nabla^2\psi(\mathbf{r}) + V(\mathbf{r})\psi(\mathbf{r}) = E\psi(\mathbf{r}), \tag{8.1}$$

where $V(\mathbf{r})$ stands for the phonon-induced attractive interaction between the electrons. To solve for the eigenenergy, we write the wave function in momentum space

$$\psi(\mathbf{r}) = \sum_{k>k_F} f_k e^{i\mathbf{k}\cdot\mathbf{r}}, \tag{8.2}$$

where the summation runs over states above the Fermi surface due to a fully filled Fermi sea. Inserting the equation above into the Schrödinger's equation and applying the orthogonality condition of the plane-wave basis, we get the eigenequation in momentum space

$$2\epsilon_k f_k + \sum_{k'>k_F} V_{k,k'} f_{k'} = E f_k, \tag{8.3}$$

where $V_{k,k'} = \int d\mathbf{r} V(\mathbf{r})e^{-i(\mathbf{k}-\mathbf{k}')\cdot\mathbf{r}}$. To simplify the problem, we assume that the interaction is attractive and constant within a thin shell of width $2\hbar\omega_c$ around the Fermi surface. This is valid when the Fermi energy E_F is much larger than the width of the shell ω_c, typically on the order of the Debye energy of the metal. Under this assumption, the interaction can be written as [3]

$$V_{k,k'} = -U\theta(|\epsilon_k - E_F| - \hbar\omega_c)\theta(|\epsilon_{k'} - E_F| - \hbar\omega_c), \tag{8.4}$$

where $\theta(x)$ is the Heaviside step function. Substituting the simplified inter-action potential into the Schrödinger's equation, we get

$$(2\epsilon_k - E)f_k = U \sum_{k'>k_F}' f_{k'}, \tag{8.5}$$

where $\sum'$ specifies the step-functions in Eq. (8.4) are satisfied during sum-mation. Assuming $2\epsilon_k \neq E$,[1] we move the factor $2\epsilon_k - E$ over and perform summation $\sum_{k'>k_F}'$ on both sides,

$$\frac{1}{U} = \sum_{k>k_F}' \frac{1}{2\epsilon_k - E}. \tag{8.6}$$

Casting the summation into integral,

$$\frac{1}{U} = \int_{E_F}^{E_F + \hbar\omega_c} \frac{d\epsilon\, g(\epsilon)}{2\epsilon - E},$$
$$\approx \frac{N_F}{2} \ln \frac{2E_F - E + 2\hbar\omega_c}{2E_F - E}. \tag{8.7}$$

Note that we have assumed a constant density of state N_F within the energy shell of interest $(E_F - \hbar\omega_c, E_F + \hbar\omega_c)$. Assuming $N_F U \ll 1$, we have

$$E \approx 2E_F - 2\hbar\omega_c \exp\left(-\frac{2}{N_F U}\right). \tag{8.8}$$

Clearly, the energy is slightly smaller than what one would have by simply adding two electrons above the Fermi surface. This implies the existence of a bound state. However, if these two electrons can form a bound state, all the other electrons near the Fermi surface should behave similarly. Hence, the true ground state in the presence of an attractive interaction, no matter how small the interaction is, is going to be drastically different from the ground state of a non-interacting Fermi gas. What then is the form of this new many-body ground state?

8.2.   BCS theory

As one of the most successful theories in physics, the BCS theory was first proposed in 1957 to explain the phenomenon of superconductivity in

---

[1]We will show $E < 2E_F$, which is consistent with the assumption here.

metals [4]. The essence of the theory lies in the prediction of the pairing of electrons with opposite momenta in the presence of an arbitrarily weak attractive interaction.

As a minimal description, we start from the BCS Hamiltonian

$$H - \mu N = \sum_{\mathbf{k},\sigma} (\epsilon_{\mathbf{k}} - \mu) a^{\dagger}_{\mathbf{k},\sigma} a_{\mathbf{k},\sigma} + \frac{1}{\mathcal{V}} \sum_{\mathbf{k},\mathbf{k}'} U_{\mathbf{k},\mathbf{k}'} a^{\dagger}_{\mathbf{k},\uparrow} a^{\dagger}_{-\mathbf{k},\downarrow} a_{-\mathbf{k}',\downarrow} a_{\mathbf{k}',\uparrow}, \qquad (8.9)$$

where $\epsilon_{\mathbf{k}} = k^2/(2m)$ (m is the electron mass and $\hbar = 1$), μ is the chemical potential, $\mathcal{V}$ is the quantization volume, $a^{\dagger}_{\mathbf{k},\sigma}$ ($a_{\mathbf{k},\sigma}$) is the creation (annihilation) operator for an electron with momentum $\mathbf{k}$, $N = \sum_{\mathbf{k}\sigma} a^{\dagger}_{\mathbf{k}\sigma} a_{\mathbf{k}\sigma}$ is the total particle number and $\sigma = \uparrow, \downarrow$ is the electronic spin. Here, the effective electron-electron coupling rate $U_{\mathbf{k},\mathbf{k}'}$ is considered to be negative and arbitrarily small. Similar to the assumption in the previous section, the BCS theory assumes that $U_{\mathbf{k},\mathbf{k}'}$ is a constant U within the range of approximately the Debye energy near the Fermi surface, and zero otherwise.

The Hamiltonian in Eq. (8.9) is not quadratic and hence cannot be diagonalized analytically. To proceed, let us recall the physical picture of Cooper instability that we encountered in the previous section. The general picture there is that for electrons moving near the Fermi surface in the presence of an attractive interaction, bound pairs with zero center-of-mass momentum tend to form between different spin species. When a large number of electrons form such pairs, the nature of the ground state naturally changes. One approach to characterize the ground state is to "guess" a physically relevant trial wave function for the ground state and then minimize the expectation value of the Hamiltonian to determine the variational parameters in the wave function. A simple but highly non-trivial guess was given by Bardeen et al. in 1957:

$$|\Psi\rangle = \prod_{\mathbf{k}} \left(u_{\mathbf{k}} + v_{\mathbf{k}} a^{\dagger}_{\mathbf{k},\uparrow} a^{\dagger}_{-\mathbf{k},\downarrow} \right) |\text{vac}\rangle, \qquad (8.10)$$

where complex numbers $u_{\mathbf{k}}$ and $v_{\mathbf{k}}$ contain information regarding the wave function of the correlated electron pairs, and we have $|u_{\mathbf{k}}|^2 + |v_{\mathbf{k}}|^2 = 1$ as required by the normalization condition $\langle \Psi | \Psi \rangle = 1$. This is the famous BCS wave function. To gain further insight of this wave function, we rewrite it as the following

$$|\Psi\rangle = \prod_{\mathbf{k}} u_{\mathbf{k}} \left(1 + f_{\mathbf{k}} a^{\dagger}_{\mathbf{k},\uparrow} a^{\dagger}_{-\mathbf{k},\downarrow}\right) |\text{vac}\rangle$$

$$= \prod_{\mathbf{k}} u_{\mathbf{k}} \prod_{\mathbf{k}} \left(1 + f_{\mathbf{k}} a^{\dagger}_{\mathbf{k},\uparrow} a^{\dagger}_{-\mathbf{k},\downarrow}\right) |\text{vac}\rangle$$

$$= \mathcal{N} \exp\left(\sum_{\mathbf{k}} f_{\mathbf{k}} a^{\dagger}_{\mathbf{k},\uparrow} a^{\dagger}_{-\mathbf{k},\downarrow}\right) |\text{vac}\rangle, \tag{8.11}$$

where the normalization constant $\mathcal{N} = \prod_{\mathbf{k}} u_{\mathbf{k}}$. Apparently, the BCS wave function is equivalent to a coherent state of Cooper pairs with the pairing wave function $f_{\mathbf{k}} = v_{\mathbf{k}}/u_{\mathbf{k}}$. As the Cooper pairs can be considered as composite bosons, the BCS state can also be viewed as a condensate of Cooper pairs, with the pairing wave function $f_{\mathbf{k}}$ carrying information regarding the correlated electron pairs. The pairing wave function $f_{\mathbf{k}}$ as well as the ground state energy can be calculated from the extrema condition of the energy expectation value $\partial\langle\Psi|H - \mu N|\Psi\rangle/\partial f_{\mathbf{k}}^* = 0$ [3].

Equivalently, one may follow the Bogoliubov's approach to diagonalize the Hamiltonian, as we will demonstrate below [5]. To diagonalize the Hamiltonian following the Bogoliubov's approach, one needs to introduce the pair operator $a_{-\mathbf{k},\downarrow} a_{\mathbf{k},\uparrow}$ and its Hermitian conjugate. It is easy to see that the expectation value of this pair operator vanishes for vacuum; however, it is non-vanishing for the BCS ground state. Being essentially a mean-field theory, the BCS theory assumes that the quantum fluctuation of the pair operator around its expectation value $\delta = a_{-\mathbf{k},\downarrow} a_{\mathbf{k},\uparrow} - \langle a_{-\mathbf{k},\downarrow} a_{\mathbf{k},\uparrow}\rangle$ is small and only needs to be retained up to the first order. This implies considerable population of the pairing state. More explicitly, we have

$$a^{\dagger}_{\mathbf{k},\uparrow} a^{\dagger}_{-\mathbf{k},\downarrow} a_{-\mathbf{k}',\downarrow} a_{\mathbf{k}',\uparrow}$$

$$= \left[\left\langle a^{\dagger}_{\mathbf{k},\uparrow} a^{\dagger}_{-\mathbf{k},\downarrow}\right\rangle + \left(a^{\dagger}_{\mathbf{k},\uparrow} a^{\dagger}_{-\mathbf{k},\downarrow} - \left\langle a^{\dagger}_{\mathbf{k},\uparrow} a^{\dagger}_{-\mathbf{k},\downarrow}\right\rangle\right)\right]$$

$$\times \left[\langle a_{-\mathbf{k}',\downarrow} a_{\mathbf{k}',\uparrow}\rangle + (a_{-\mathbf{k}',\downarrow} a_{\mathbf{k}',\uparrow} - \langle a_{-\mathbf{k}',\downarrow} a_{\mathbf{k}',\uparrow}\rangle)\right]$$

$$= \left\langle a^{\dagger}_{\mathbf{k},\uparrow} a^{\dagger}_{-\mathbf{k},\downarrow}\right\rangle a_{-\mathbf{k}',\downarrow} a_{\mathbf{k}',\uparrow} + a^{\dagger}_{\mathbf{k},\uparrow} a^{\dagger}_{-\mathbf{k},\downarrow} \langle a_{-\mathbf{k}',\downarrow} a_{\mathbf{k}',\uparrow}\rangle$$

$$- \left\langle a^{\dagger}_{\mathbf{k},\uparrow} a^{\dagger}_{-\mathbf{k},\downarrow}\right\rangle \langle a_{-\mathbf{k}',\downarrow} a_{\mathbf{k}',\uparrow}\rangle + O(\delta^2). \tag{8.12}$$

Defining a gap function $\Delta = -U/\mathcal{V} \sum_{\mathbf{k}}' \langle a_{-\mathbf{k},\downarrow} a_{\mathbf{k},\uparrow} \rangle$, whose physical significance will become clear shortly, one gets the effective Hamiltonian

$$H - \mu N = \sum_{\mathbf{k},\sigma} (\epsilon_{\mathbf{k}} - \mu) a_{\mathbf{k},\sigma}^{\dagger} a_{\mathbf{k},\sigma} - \sum_{\mathbf{k}} \left(\Delta a_{\mathbf{k},\uparrow}^{\dagger} a_{-\mathbf{k},\downarrow}^{\dagger} + \Delta^{*} a_{-\mathbf{k},\downarrow} a_{\mathbf{k},\uparrow} \right)$$

$$- \mathcal{V} |\Delta|^2 / U, \tag{8.13}$$

where the $\sum_{\mathbf{k}}'$ in the definition of the gap function Δ implies that the summation is only taken within a thin shell around the Fermi surface, as we have discussed before.

The effective Hamiltonian is now quadratic and can be diagonalized exactly. We observe that field operators in the subspace spanned by $\{\mathbf{k} \uparrow, -\mathbf{k} \downarrow\}$ only couple with those in the same subspace. In other words, the effective Hamiltonian is block diagonal when we group the eigenbases accordingly. One can show this more explicitly by writing

$$H_{\text{eff}} = \sum_{\mathbf{k}} \begin{pmatrix} a_{\mathbf{k},\uparrow}^{\dagger} & a_{-\mathbf{k},\downarrow} \end{pmatrix} \begin{pmatrix} \epsilon_{\mathbf{k}} - \mu & -\Delta \\ -\Delta^{*} & -(\epsilon_{\mathbf{k}} - \mu) \end{pmatrix} \begin{pmatrix} a_{\mathbf{k},\uparrow} \\ a_{-\mathbf{k},\downarrow}^{\dagger} \end{pmatrix}$$

$$+ \sum_{\mathbf{k}} (\epsilon_{\mathbf{k}} - \mu) - \frac{|\Delta|^2}{U}. \tag{8.14}$$

The problem is then reduced to diagonalizing a 2×2 matrix. To make the process more transparent, we apply the Bogoliubov transformation to mix electrons with holes

$$\alpha_{\mathbf{k},\uparrow} = u_{\mathbf{k}}^{*} a_{\mathbf{k},\uparrow} - v_{\mathbf{k}}^{*} a_{-\mathbf{k},\downarrow}^{\dagger}, \tag{8.15}$$

$$\alpha_{\mathbf{k},\downarrow}^{\dagger} = v_{\mathbf{k}} a_{\mathbf{k},\uparrow} + u_{\mathbf{k}} a_{-\mathbf{k},\downarrow}^{\dagger}, \tag{8.16}$$

where $u_{\mathbf{k}}$ and $v_{\mathbf{k}}$ are complex numbers. To ensure that the transformation is canonical, i.e., the operators $\alpha_{\mathbf{k},\sigma}$ also satisfy the Fermi–Dirac anticommutation relations, we require $|u_{\mathbf{k}}|^2 + |v_{\mathbf{k}}|^2 = 1$. Substituting these into the effective Hamiltonian and requiring that the resulting Hamiltonian be diagonal in $\alpha_{\mathbf{k},\sigma}$, we get a set of equations for $u_{\mathbf{k}}$ and $v_{\mathbf{k}}$

$$2 v_{\mathbf{k}} u_{\mathbf{k}} \xi_{\mathbf{k}} + \Delta v_{\mathbf{k}}^2 - \Delta^{*} u_{\mathbf{k}}^2 = 0, \tag{8.17}$$

$$|v_{\mathbf{k}}|^2 + |u_{\mathbf{k}}|^2 = 1, \tag{8.18}$$

where $\xi_{\mathbf{k}} = \epsilon_{\mathbf{k}} - \mu$. As only the relative phase between $v_{\mathbf{k}}$ and $u_{\mathbf{k}}$ is important, let us assume that $u_{\mathbf{k}}$ be real. Dividing $u_{\mathbf{k}}^2$ on both sides of Eq. (8.17), we get an equation for $f_{\mathbf{k}} = v_{\mathbf{k}}/u_{\mathbf{k}}$

$$\Delta f_{\mathbf{k}}^2 + 2\xi_{\mathbf{k}} f_{\mathbf{k}} - \Delta^* = 0, \tag{8.19}$$

which leads to the solution $f_{\mathbf{k}}^{\pm} = (-\xi_{\mathbf{k}} \pm E_{\mathbf{k}})/\Delta$, with $E_{\mathbf{k}} = \sqrt{\xi_{\mathbf{k}}^2 + |\Delta|^2}$. It is easy to see that $f_{\mathbf{k}}^{-}$ corresponds to quasi-hole excitations, i.e., with negative excitation spectrum $-E_{\mathbf{k}}$. Hence, we only consider the $f_{\mathbf{k}}^{+}$ branch and omit the superscript henceforth. Together with the normalization condition, we have

$$u_{\mathbf{k}}^2 = \frac{[E_{\mathbf{k}} + (\epsilon_{\mathbf{k}} - \mu)]}{E_{\mathbf{k}}},$$

$$v_{\mathbf{k}}^2 = \frac{[E_{\mathbf{k}} - (\epsilon_{\mathbf{k}} - \mu)]}{E_{\mathbf{k}}}. \tag{8.20}$$

The physical interpretation of $E_{\mathbf{k}}$ is obtained by looking at the diagonalized effective Hamiltonian

$$H - \mu N = \sum_{\mathbf{k}} E_{\mathbf{k}} (\alpha_{\mathbf{k},\uparrow}^{\dagger} \alpha_{\mathbf{k},\uparrow} + \alpha_{\mathbf{k},\downarrow}^{\dagger} \alpha_{\mathbf{k},\downarrow})$$

$$+ \sum_{\mathbf{k}} [(\epsilon_{\mathbf{k}} - \mu) - E_{\mathbf{k}}] - \mathcal{V}|\Delta|^2/U. \tag{8.21}$$

The expectation value of the first term on the right-hand side is positive semidefinite. Therefore, the extrema condition for the energy expectation $\langle H - \mu N \rangle$ requires that the ground state be the vacuum of the operators $\alpha_{\mathbf{k},\sigma}$, i.e., $\alpha_{\mathbf{k},\sigma}|G\rangle = 0, \forall \mathbf{k}, \sigma$. One can easily check that the BCS state $|\Psi\rangle$ satisfies this condition, so long as $v_{\mathbf{k}}/u_{\mathbf{k}} = f_{\mathbf{k}}$. In this sense, the approach that we take is equivalent to the variational functional approach starting with the BCS wave function, with the understanding that the operators $\alpha_{\mathbf{k},\sigma}$ describe quasi-particle excitations above the BCS ground state with the dispersion spectrum $E_{\mathbf{k}}$. From Fig. 8.1(a), we see that the gap function Δ coincides with the minimum of the excitation spectrum. This implies that the minimum energy required to break up a Cooper pair and create two quasi-particle excitations is 2Δ. The ground state energy can be read off the effective Hamiltonian in Eq. (8.21) by setting the first term on the right-hand side to be zero.

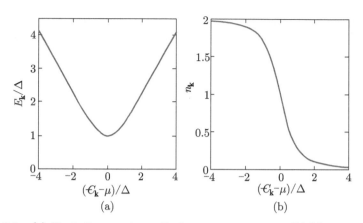

Figure 8.1. (a) Excitation spectrum $E_{\mathbf{k}}$ in momentum space. (b) Momentum space distribution $n_{\mathbf{k}}$ at zero temperature. The results are from a BCS mean-field calculation.

Of course, we still need to solve for the gap function Δ. The self-consistent equation for Δ can be extracted from its definition

$$\Delta = -\frac{U}{\mathcal{V}} \sideset{}{'}\sum_{\mathbf{k}} \langle a_{-\mathbf{k},\downarrow} a_{\mathbf{k},\uparrow} \rangle$$

$$= -\frac{U}{\mathcal{V}} \sideset{}{'}\sum_{\mathbf{k}} \left\langle \left(u_{\mathbf{k}}^{*} \alpha_{\mathbf{k},\downarrow} - v_{\mathbf{k}} \alpha_{\mathbf{k},\uparrow}^{\dagger} \right) \left(u_{\mathbf{k}}^{*} \alpha_{\mathbf{k},\uparrow} + v_{\mathbf{k}} \alpha_{\mathbf{k},\downarrow}^{\dagger} \right) \right\rangle$$

$$= -\frac{U}{\mathcal{V}} \sideset{}{'}\sum_{\mathbf{k}} \left(u_{\mathbf{k}}^{*} v_{\mathbf{k}} \left\langle \alpha_{\mathbf{k},\downarrow} \alpha_{\mathbf{k},\downarrow}^{\dagger} \right\rangle - v_{\mathbf{k}} u_{\mathbf{k}}^{*} \left\langle \alpha_{\mathbf{k},\uparrow}^{\dagger} \alpha_{\mathbf{k},\uparrow} \right\rangle \right). \qquad (8.22)$$

At zero temperature, we may take $\left\langle \alpha_{\mathbf{k},\sigma}^{\dagger} \alpha_{\mathbf{k},\sigma} \right\rangle = 0$, as the expectation value is taken with respect to the ground state. The resulting equality is the well-known gap equation at zero temperature

$$\frac{1}{U} = -\frac{1}{\mathcal{V}} \sideset{}{'}\sum_{\mathbf{k}} \frac{1}{2E_{\mathbf{k}}}. \qquad (8.23)$$

The summation in this gap equation does not diverge due to the finite range in the summation over $\mathbf{k}$.

Note that in the weak-coupling limit of the standard BCS theory, the Fermi surface is only slightly perturbed. Hence the chemical potential can be well approximated by the Fermi energy. With this understanding, the gap equation above can be analytically solved in a similar fashion as that of the bound state energy in the previous section

$$
1 = \frac{|U|}{\mathcal{V}} \sum_{\mathbf{k}} \frac{1}{2E_{\mathbf{k}}}
$$

$$
= \int_{E_F - \hbar\omega_c}^{E_F + \hbar\omega_c} d\epsilon \frac{g(\epsilon)}{2} \frac{|U|}{2\sqrt{(\epsilon - E_F)^2 + |\Delta|^2}}
$$

$$
\approx |U| \int_0^{\hbar\omega_c/\Delta} dx \frac{N_F}{2\sqrt{x^2 + 1}}, \quad x = \epsilon - E_F
$$

$$
= \frac{|U|N_F}{2} \sinh^{-1} \frac{\hbar\omega_c}{\Delta}, \tag{8.24}
$$

where we have assumed that the density of state $g(\epsilon)$ changes very little in the range of integration and which we have approximated as a constant N_F. From the derivation above, we can immediately write down the zero-temperature gap function

$$
\Delta \approx 2\hbar\omega_c \exp\left[-\frac{2}{g(E_F)|U|}\right], \tag{8.25}
$$

which implies that the gap function remains finite no matter how small the interaction rate $|U|$ is and that it scales exponentially with the interaction rate $|U|$ in the weak-coupling limit.

At a finite temperature T, the term $\langle \alpha_{\mathbf{k},\sigma}^{\dagger} \alpha_{\mathbf{k},\sigma} \rangle$ is no longer zero. By definition, this term is the density distribution of quasi-particle excitations. We may assume that these fermionic particles obey the Fermi–Dirac distribution so that the finite temperature gap equation takes the form

$$
-\frac{1}{U} = \frac{1}{\mathcal{V}} \sum_{\mathbf{k}} \frac{1 - 2f(E_k)}{2E_k}, \tag{8.26}
$$

with $f(x) = 1/(e^{\beta x} + 1)$. Note that the chemical potential of the quasi-particle excitations is zero as the number of these excitations is not conserved.

The critical temperature for the BCS superfluidity can be calculated from the finite temperature gap equation by setting $\Delta = 0$

$$1 = \frac{|U|}{\mathcal{V}} \sum_{\mathbf{k}} \frac{1 - 2f(E_k)}{2E_k} \Bigg|_{T=T_c, \Delta=0}$$

$$\approx \int_0^{\beta_c \hbar \omega} |U| \frac{N_F}{2} \frac{dx}{x} \left[1 - \frac{2}{e^x + 1} \right], \quad x = \beta_c(\epsilon_k - \mu)$$

$$= \frac{|U| g(\epsilon_F)}{2} \left(\ln \beta_c \hbar \omega + C \right), \tag{8.27}$$

with the constant $C = \ln\left(\frac{2\gamma_E}{\pi}\right) \approx -1$ and the Euler constant $\gamma_E \approx 0.577$. Therefore

$$k_B T_c \approx -\hbar \omega \exp\left[-\frac{2}{|U| N_F} \right]. \tag{8.28}$$

With modifications considering realistic electron-phonon interaction potentials, many experimental results can be explained surprisingly well with this simple mean-field theory. This suggests that the experiments are essentially in the weakly-coupling limit.

Finally, the perturbation of the Fermi surface can be vividly demonstrated by the momentum-space density distribution

$$n_{\mathbf{k}} = \sum_{\sigma} \langle a_{\mathbf{k},\sigma}^\dagger a_{\mathbf{k},\sigma} \rangle$$

$$= 2v_{\mathbf{k}}^2, \tag{8.29}$$

which we have plotted in Fig. 8.1. The occupation profile is reminiscent of the Fermi surface for a free electron gas, but is now perturbed near the Fermi surface. The perturbation of the Fermi surface occurs over the range of Δ, which is exponentially small in the weak-coupling limit. We will see in the next section how these features will change throughout the BCS-BEC crossover.

8.3.   Description of BCS-BEC crossover on the mean-field level

The theoretical study of the BCS-BEC crossover dates back to the seminal works of Eagles (1969) and Leggett (1980), where they pointed out that the BCS wave function can be extended to describe the physics beyond the weak-coupling limit [6, 7]. Despite being a mean-field theory, the approach paints a qualitatively correct picture throughout the BCS-BEC crossover at zero temperature, during which the s-wave scattering length a_s between different fermions can become infinitely large. With the experimental achievements of ultracold Fermi gases, especially the Fermi condensate, the study of the BCS-BEC crossover has attracted much attention in recent years in this new context. Here, we will give an outline of the BCS-BEC crossover based on the mean-field BCS theory in the previous section. We will discuss the crossover theory in the context of ultracold Fermi gases and emphasize the differences with respect to the standard BCS theory. Although the zero-temperature theory in this section can be easily extended to finite temperatures, the large thermal fluctuations make the mean-field theory at finite temperatures less satisfying.

The primary assumption of the mean-field crossover theory is that even away from the weak-coupling region, the BCS wave function

$$|\Psi\rangle = \prod_{\mathbf{k}} \left(u_{\mathbf{k}} + v_{\mathbf{k}} a_{\mathbf{k},\uparrow}^{\dagger} a_{-\mathbf{k},\downarrow}^{\dagger} \right) |\text{vac}\rangle, \tag{8.30}$$

can still apply, provided that the chemical potential of the fermions be determined self-consistently from the number constraints as the interaction rate is changed.

The BCS-BEC crossover in ultracold atoms can be modeled by the Hamiltonian

$$H - \mu N = \sum_{\mathbf{k},\sigma} (\epsilon_{\mathbf{k}} - \mu) a_{\mathbf{k},\sigma}^{\dagger} a_{\mathbf{k},\sigma} + \frac{1}{\mathcal{V}} \sum_{\mathbf{k},\mathbf{k}'} U_{\mathbf{k},\mathbf{k}'} a_{\mathbf{k},\uparrow}^{\dagger} a_{-\mathbf{k},\downarrow}^{\dagger} a_{-\mathbf{k}',\downarrow} a_{\mathbf{k}',\uparrow}, \tag{8.31}$$

which is similar to the BCS Hamiltonian. The important difference lies in the origin and the form of the interaction rate $U_{\mathbf{k},\mathbf{k}'}$. As discussed in Chap. 5, in the case that the effective range r_0 of the two-body s-wave interaction is much smaller than all the other relevant length scales in the system, the effects of the s-wave interaction can be captured by a contact potential $U(\mathbf{r} - \mathbf{r}') = 4\pi\hbar^2 a_s \delta(\mathbf{r} - \mathbf{r}')/m$. The ultraviolet divergence of the contact potential, which originates from the unphysical behavior of the contact potential at high energies, can be renormalized by requiring that the Hamiltonian with the effective contact interaction reproduces the same two-body physics. In the spirit of this, we transform the contact potential into momentum space and write the interaction rate $U_{\mathbf{k},\mathbf{k}'}$ as a bare constant

U, up to a very large momentum cutoff k_c that is related to the effective range r_0.

Following the standard renormalization procedure of the contact potentials detailed in Chap. 5, the bare interaction rate U can be connected with the physical one $U_p = 4\pi\hbar^2 a_s/m$ via the renormalization relation in three dimensions

$$\frac{1}{U} = \frac{1}{U_p} - \sum_{\mathbf{k}} \frac{1}{2\epsilon_{\mathbf{k}}}. \tag{8.32}$$

The renormalized gap equation then becomes

$$-\frac{1}{U_p} = \sum_{\mathbf{k}} \left(\frac{1}{2E_{\mathbf{k}}} - \frac{1}{2\epsilon_{\mathbf{k}}} \right). \tag{8.33}$$

Compared to the gap equation in Eq. (8.23), the summation is now over the entire momentum space and is convergent with a properly renormalized bare interaction rate U. The large momentum cutoff k_c also drops out of the problem during the renormalization.

Beyond the weak-coupling limit, pairing may significantly affect the shape of the Fermi surface. Therefore, the chemical potential can no longer be simply approximated by the Fermi energy. Instead, we need to solve the chemical potential μ self-consistently from the number constraints $N = -\partial\Omega/\partial\mu$, which gives the number equation

$$n = \frac{1}{\mathcal{V}} \sum_{\mathbf{k}} \left(1 - \frac{\epsilon_k - \mu}{E_k} \right). \tag{8.34}$$

For each different scattering length, we must solve Eqs. (8.33) and (8.34) self-consistently to determine the values of the gap function Δ and the chemical potential μ.

In three dimensions, it has been shown that Eqs. (8.33) and (8.34) can be solved analytically in terms of elliptical integrals [8]. Employing the relations $E_F = \hbar^2 k_F^2/2m$ and $k_F^3 = 3\pi^2 n$, it is possible to cast the equations into dimensionless forms [9]

$$\frac{\Delta}{E_F} = \left(\frac{2}{3I_2(\frac{\mu}{\Delta})} \right)^{\frac{2}{3}}, \tag{8.35}$$

$$\frac{1}{k_F a_s} = -\frac{2}{\pi} \left(\frac{2}{3I_2(\frac{\mu}{\Delta})} \right)^{\frac{1}{3}} I_1\left(\frac{\mu}{\Delta} \right), \tag{8.36}$$

where the dimensionless integrals $I_1(z)$ and $I_2(z)$ are defined as

$$I_1(z) = \int_0^\infty dx x^2 \left(\frac{1}{\sqrt{(x^2 - z)^2 + 1}} - \frac{1}{x^2} \right), \tag{8.37}$$

$$I_2(z) = \int_0^\infty dx x^2 \left(1 - \frac{x^2 - z}{\sqrt{(x^2 - z)^2 + 1}} \right). \tag{8.38}$$

Let us first examine the two limiting cases. For $(k_F a_s)^{-1} \to -\infty$, Δ/E_F becomes exponentially small and μ approaches the Fermi energy $E_F = \hbar^2 k_F^2 / 2m$, which implies that the system is in the BCS limit. From the dispersion $E_\mathbf{k} = \sqrt{(\epsilon_\mathbf{k} - \mu)^2 + \Delta^2}$, we see that the minimum of the excitation gap is the same as the order parameter Δ. From the analytical expressions Eqs. (8.35) and (8.36), one may take the limit $z \to +\infty$ and get

$$\mu \approx E_F, \tag{8.39}$$

$$\Delta \approx 8e^{-2} \exp \left(\frac{\pi}{2k_F a_s} \right). \tag{8.40}$$

These relations tell us that in the BCS limit, the physical picture is similar to what we discussed in the previous section for the standard BCS theory. Although the origin and the form of the interactions are different, the fermions form Cooper pairs with large spatial correlation lengths in the weak-coupling limit.

For $(k_F a_s)^{-1} \to 0^+$, Δ/E_F becomes very large and μ becomes large and negative. Here the system is in the BEC limit and the minimum of the excitation spectrum is given by $\sqrt{\mu^2 + \Delta^2}$, which is roughly half of the minimum energy required to break a molecular bound state in this limit. In the analytical expressions Eqs. (8.35) and (8.36), one may take the limit $z \to -\infty$ and get

$$\mu \approx -\frac{\hbar^2}{2m a_s} + \frac{\pi \hbar^2 a_s n}{m}, \tag{8.41}$$

$$\Delta \approx \sqrt{\frac{16}{3\pi k_F a_s}} E_F. \tag{8.42}$$

The first term in the expression for the chemical potential corresponds to one half of the binding energy of the molecules and the second term stands for the interaction energy, which is repulsive here. As the chemical potential of the molecules μ_m is twice of that of the fermions in Eq. (8.41), we may rewrite the second term as

$$\mu_m = \frac{4\pi\hbar^2 a_m(\frac{n}{2})}{2m} = 2\frac{\pi\hbar^2 a_s n}{m}, \qquad (8.43)$$

so that we may identify the scattering length between the molecules to be $a_m = 2a_s$. This mean-field result proved to be inaccurate, as the correlations between fermions have not been considered. An exact result can be obtained by considering a four-body scattering problem, which gives $a_m = 0.6a_s$ [10]. Therefore, in the BEC limit, we can think of the Fermi gas as a bunch of weakly interacting composite bosons. This understanding, together with the molecular scattering length $a_m = 0.6a_s$, agrees with Quantum Monte Carlo simulations [11].

In between these two limiting cases, the system changes smoothly from a BCS-like many-body state with long-range correlations to a BEC of weakly interacting bound molecular dimers. In Fig. 8.2, we solve Eqs. (8.33) and (8.34) numerically and plot the variation of Δ and μ as functions of $(k_F a_s)^{-1}$, where k_F is the Fermi wave vector determined from the total density $k_F = (3\pi^2 n)^{1/3}$. The chemical potential decreases from E_F in the BCS limit and crosses zero on the BEC side of the resonance, when the pairing order increases toward resonance and becomes comparable to the Fermi energy in the strongly interacting region $-1 < (k_F a_s)^{-1} < 1$. One can also see from the momentum space density distribution that the perturbation near the Fermi surface in the BCS limit grows as the interaction rate

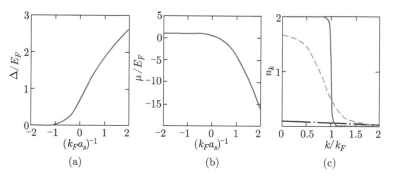

(a) (b) (c)

Figure 8.2. Properties throughout the BCS-BEC crossover. (a) The pairing gap Δ as a function of $(k_F a_s)^{-1}$. (b) The chemical potential as a function of $(k_F a_s)^{-1}$. (c) The momentum space density distribution for various interaction strengths: $(k_F a_s)^{-1} = -1$ (solid), $(k_F a_s)^{-1} = 0$ (dashed) and $(k_F a_s)^{-1} = 1$ (dash-dotted).

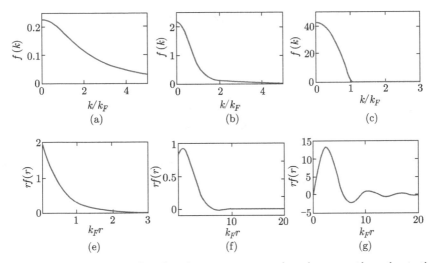

Figure 8.3.   Pairing wavefunction in momentum and real spaces throughout the crossover: (a) and (e) $(k_F a_s)^{-1} = 1$; (b) and (f) $(k_F a_s)^{-1} = 0$, and (c) and (g) $(k_F a_s)^{-1} = -1$.

increases. Eventually, the Fermi surface becomes hardly discernible on the BEC side of the resonance. Throughout the process, all physical quantities change smoothly and there is no change of symmetry. Therefore, the system does not go through any phase transitions. In Fig. 8.3, we show the pairing wave function $f_\mathbf{k} = v_\mathbf{k}/u_\mathbf{k}$ and its Fourier transform $f(\mathbf{r}) = \int f(\mathbf{k})e^{i\mathbf{k}\cdot\mathbf{r}}d\mathbf{k}$ throughout the crossover region. We can clearly see how the size of the pairing states smoothly varies from very large to very small between the two limiting cases. Importantly, the size of a fermion pair at resonance is roughly on the order of the inter-particle separation. We also note that the mean-field theory applied to get these results is not quantitatively accurate in the strongly interacting region. However, it still provides a qualitatively correct physical picture.

The formalism above can be easily extended to the finite-temperature case at the mean-field level. For concreteness, we will discuss only the finite-temperature formalism for a single-channel model in the absence of a trapping potential. Finite-temperature schemes for other variations can be derived similarly.

At finite temperatures, the partition function in the grand canonical ensemble is

$$Z = \text{tr}\, e^{-\beta[H-\mu N]}, \tag{8.44}$$

where $\beta = 1/k_B T$ and k_B is the Boltzmann constant. The thermodynamic potential can be written as

$$\Omega = -\frac{1}{\beta} \ln Z = -\frac{1}{\beta} \ln \mathrm{tr}\, e^{-\beta[H-\mu N]}. \qquad (8.45)$$

A closed-form expression for the thermodynamic potential does not exist in general except on the mean-field level, where we may replace $H - \mu N$ in the exponential with the diagonalized effective Hamiltonian Eq. (8.21). We then take the trace in the basis of the eigenmodes of the quasi-particle excitations, which leads to

$$\Omega = -\frac{|\Delta|^2}{U} - \frac{1}{\beta} \sum_{\mathbf{k}} \{2 \ln [1 + \exp(-\beta E_k)] - \beta(\epsilon_k - \mu - E_k)\}. \qquad (8.46)$$

The finite-temperature version of the gap equation and number equation can be obtained following the extrema condition $\partial \Omega / \partial \Delta^* = 0$ and the number constraint $N = -\partial \Omega / \partial \mu$

$$-\frac{1}{U} = \frac{1}{V} \sum_{\mathbf{k}} \left(\frac{1 - 2f(E_k)}{2E_k} \right) = \frac{1}{V} \sum_{\mathbf{k}} \frac{1}{2E_k} \tanh \left(\frac{\beta E_k}{2} \right), \qquad (8.47)$$

$$n = \frac{1}{V} \sum_{\mathbf{k}} \left[1 - \frac{\epsilon_k - \mu}{E_k} + 2\frac{\epsilon_k - \mu}{E_k} f(E_k) \right]$$

$$= \frac{1}{V} \sum_{\mathbf{k}} \left[1 - \frac{\epsilon_k - \mu}{E_k} \tanh \left(\frac{\beta E_k}{2} \right) \right]. \qquad (8.48)$$

As in the zero-temperature case, these equations can be solved self-consistently to determine the pairing order and chemical potential at finite temperatures.

8.4. Feshbach resonance and the two-channel model

The ultracold Fermi gas is an ideal system to study the BCS-BEC crossover physics in experiment. This is made possible by the employment of the Feshbach resonance technique, which can be described by a resonant scattering process between two scattering channels, the closed channel and the open channel. The most natural model therefore should include both scattering channels. The Hamiltonian Eq. (8.31), known as the single-channel model, only contains physics in the open channel. It is interesting to investigate the validity of the single-channel Hamiltonian in describing the crossover physics in the experiments with Feshbach resonances. As we will see later, the single-channel model can already account for most of the interesting physics near a so-called wide Feshbach resonance, which is the most commonly studied scenari experimentally.

The two-channel Hamiltonian was first adopted in the study of the BCS-BEC crossover problem by M. Holland et al. [12]. The closed-channel bosons, which are essentially quasi-bound molecules of the high-lying vibrational states, are modeled using structureless bosons

$$H - \mu N = \sum_{\sigma, \mathbf{k}} (\epsilon_{\mathbf{k}} - \mu) a_{\mathbf{k}, \sigma}^{\dagger} a_{\mathbf{k}, \sigma} + \sum_{\mathbf{q}} (\gamma + \epsilon_{\mathbf{q}}/2 - 2\mu) b_{\mathbf{q}}^{\dagger} b_{\mathbf{q}}$$

$$+ \frac{g}{\sqrt{\mathcal{V}}} \sum_{\mathbf{k}, \mathbf{q}} \left(b_{\mathbf{q}}^{\dagger} a_{-\mathbf{k}+\mathbf{q}/2, \downarrow} a_{\mathbf{k}+\mathbf{q}/2, \uparrow} + H.C. \right)$$

$$+ \frac{U}{\mathcal{V}} \sum_{\mathbf{k}, \mathbf{k}', \mathbf{q}} a_{\mathbf{k}+\mathbf{q}/2, \uparrow}^{\dagger} a_{-\mathbf{k}+\mathbf{q}/2, \downarrow}^{\dagger} a_{-\mathbf{k}'+\mathbf{q}/2, \downarrow} a_{\mathbf{k}'+\mathbf{q}/2, \uparrow}, \quad (8.49)$$

where $a_{\mathbf{k}}^{\dagger}, b_{\mathbf{q}}^{\dagger}$ are the creation operators for the open-channel fermions and the closed-channel bosons, respectively. The total particle number is now $N = \sum_{\mathbf{k}\sigma} a_{\mathbf{k}\sigma}^{\dagger} a_{\mathbf{k}\sigma} + 2\sum_{\mathbf{q}} b_{\mathbf{q}}^{\dagger} b_{\mathbf{q}}$. The bare detuning γ denotes the energy offset of the closed-channel molecules relative to the open-channel continuum threshold. The second line describes the coupling between closed-channel molecules and open-channel fermions, with the bare coupling rate g. The last line gives the background scattering between fermions in the open channel.

To make connection between the bare parameters in the Hamiltonian and the physical parameters, one should consider a two-body scattering problem and write down the Lippmann–Schwinger equation as done previously for the single-channel model. For a two-channel model, this is not straightforward. It is however possible to demonstrate that the single-channel Lippmann–Schwinger equation still holds for the two-channel model, provided we replace the interaction potential with an effective one. This can be shown, for example, by starting from the partition function of the two-channel model and integrating out the bosonic degrees of freedom [13]. Here, we take a simpler route by adiabatically eliminating the bosonic modes.

In the Heisenberg picture, the equation of motion for the boson operator is

$$i\hbar \dot{b}_{\mathbf{q}} = [b_{\mathbf{q}}, H]$$

$$= \left(\frac{\epsilon_{\mathbf{q}}}{2} + \gamma - 2\mu\right) b_{\mathbf{q}} + g \sum_{\mathbf{k}} a_{\frac{\mathbf{q}}{2}-\mathbf{k},\downarrow} a_{\frac{\mathbf{q}}{2}+\mathbf{k},\uparrow}. \qquad (8.50)$$

Assuming $\dot{b}_{\mathbf{q}} = 0$, the bosonic mode operators can be expressed in terms of the fermionic creation (annihilation) operators

$$b_{\mathbf{q}} = -\frac{g}{\frac{\epsilon_{\mathbf{q}}}{2} + \gamma - 2\mu} \sum_{\mathbf{k}} a_{\frac{\mathbf{q}}{2}-\mathbf{k},\downarrow} a_{\frac{\mathbf{q}}{2}+\mathbf{k},\uparrow}. \qquad (8.51)$$

Substituting the molecular creation (annihilation) operators into the two-channel Hamiltonian Eq. (8.49), we get an effective single-channel Hamiltonian

$$H_{\text{eff}} - \mu N = \sum_{\mathbf{k},\sigma} (\epsilon_{\mathbf{k}} - \mu) a_{\mathbf{k},\sigma}^{\dagger} a_{\mathbf{k},\sigma}$$

$$+ \frac{1}{\mathcal{V}} \sum_{\mathbf{q},\mathbf{k},\mathbf{k}'} U_{\text{eff}}(\mathbf{q}) a_{\frac{\mathbf{q}}{2}+\mathbf{k},\uparrow}^{\dagger} a_{\frac{\mathbf{q}}{2}-\mathbf{k},\downarrow}^{\dagger} a_{\frac{\mathbf{q}}{2}-\mathbf{k}',\downarrow} a_{\frac{\mathbf{q}}{2}+\mathbf{k}',\uparrow}, \qquad (8.52)$$

with the effective interaction rate

$$U_{\text{eff}} = U - \frac{g^2}{\frac{\epsilon_{\mathbf{q}}}{2} + \gamma - 2\mu}. \qquad (8.53)$$

It is then easy to derive the Lippmann–Schwinger equation for this effective single-channel model

$$T = U + U \frac{1}{E - H_0} T$$

$$\Rightarrow \frac{m}{4\pi\hbar^2 a_s} = \frac{1}{U_{\text{eff}}} + \sum_{\mathbf{k}} \frac{1}{2\epsilon_{\mathbf{k}}}. \qquad (8.54)$$

For $\gamma \to \infty$, the energy offset of the molecular levels are too large for the molecules to be of any physical relevance. Therefore, the system can be described by an effective single-channel Hamiltonian, and the renormalization for the open-channel interaction rate U is

$$\frac{1}{U} = \frac{1}{U_p} - \sum_{\mathbf{k}} \frac{1}{2\epsilon_{\mathbf{k}}}, \tag{8.55}$$

with $U_p = 4\pi\hbar^2 a_{bg}/m$. For conciseness, following Ref. [14], we define $U_c^{-1} = -\sum_{\mathbf{k}} \frac{1}{2\epsilon_{\mathbf{k}}}$, so that

$$U = (1 + U_p U_c^{-1})U_p \equiv \Gamma U_p. \tag{8.56}$$

Therefore, we have renormalized the bare open-channel interaction rate in Hamiltonian Eq. (8.49). We still need to fix the renormalization condition for the bare parameters γ and g. Now, from the Lippmann–Schwinger equation of the effective Hamiltonian Eq. (8.52), we expect to find a renormalization relation

$$\frac{m}{4\pi\hbar^2 a_s} = \frac{1}{U_{\text{eff}}} - \frac{1}{U_c}. \tag{8.57}$$

It can be shown that this requires the following renormalization relations on the bare parameters [14]

$$U = \Gamma U_p, \; g = \Gamma g_p, \; \gamma = \gamma_p - \Gamma \frac{g_p^2}{U_c}. \tag{8.58}$$

Indeed, it is straightforward to check that:

$$\frac{1}{U - \frac{g^2}{\gamma - 2\mu}} - U_c^{-1} = \frac{1}{\Gamma U_p - \frac{\Gamma^2 g_p^2}{\gamma_0 - \Gamma g_p^2 U_c^{-1} - 2\mu}} - U_c^{-1}$$

$$= \frac{\gamma_p - \Gamma g_p^2 U_c^{-1} - 2\mu}{\Gamma U_p(\gamma_p - \Gamma g_p^2 U_c^{-1} - 2\mu) - \Gamma^2 g_p^2} - U_c^{-1}$$

$$= \frac{\gamma_p - \Gamma g_p^2 U_c^{-1} - 2\mu}{\Gamma U_p \gamma_p - \Gamma g_p^2 - 2\Gamma U_p \mu} - U_c^{-1}$$

$$= \Gamma^{-1} \frac{1}{U_p - \frac{g_p^2}{\gamma_p - 2\mu}} - \left(\frac{g_p^2}{U_p \gamma_p - g_p^2 - 2U_p \mu} + 1 \right) U_c^{-1}$$

$$= (\Gamma^{-1} - U_p U_c^{-1}) \frac{1}{U_p - \frac{g_p^2}{\gamma_p - 2\mu}}$$

$$= \frac{1}{U_p - \frac{g_p^2}{\gamma_p - 2\mu}},$$

which reproduces Eq. (8.56). From the two-body description of the Feshbach resonance, the scattering length can be expressed as

$$a_s = a_{bg} \left(1 - \frac{W}{B - B_0} \right), \tag{8.59}$$

where W is the width of the resonance and $B - B_0$ is the magnetic field detuning relative to the resonance point B_0. Recalling the Lippmann–Schwinger equation Eq. (8.57), we have

$$U_p = \frac{4\pi\hbar^2 a_{bg}}{m}, \quad \gamma_p = \mu_{co}(B - B_0), \quad g_p^2 = \mu_{co} W U_p, \tag{8.60}$$

where μ_{co} denotes the difference in the magnetic moment of the open- and closed-channel states.

Now that the bare parameters in the two-channel Hamiltonian are fixed, we may develop a mean-field theory similar to that for the single-channel Hamiltonian. We are interested in the low-temperature, low-energy physics, for which we may assume that the population in the pairing states with non-zero center-of-mass momenta is negligibly small. This amounts to letting $\mathbf{q} = 0$ in the Hamiltonian Eq. (8.52). Correspondingly, the effective interaction rate U_{eff} becomes independent of $\mathbf{q}$. To make the Hamiltonian quadratic, we take the mean field

$$\Delta = \frac{g}{\mathcal{V}} \langle b_0 \rangle + \frac{U}{\mathcal{V}} \sum_{\mathbf{k}} \langle a_{-\mathbf{k},\downarrow} a_{\mathbf{k},\uparrow} \rangle. \tag{8.61}$$

Equivalently, we may start from the modified BCS wave function

$$|\Psi\rangle = \mathcal{N} \exp\left(c b_0 + \sum_{\mathbf{k}} a_{\mathbf{k},\uparrow}^\dagger a_{-\mathbf{k},\downarrow}^\dagger \right) |\text{vac}\rangle, \tag{8.62}$$

which is a coherent state of the superposition of fermion pairs in the open-channel and molecules in the closed-channel.

Following Eq. (8.61), the quadratic Hamiltonian becomes

$$H_{\text{eff}} - \mu N = \sum_{\mathbf{k},\sigma} (\epsilon_{\mathbf{k}} - \mu) a_{\mathbf{k},\sigma}^\dagger a_{\mathbf{k},\sigma}$$

$$- \sum_{\mathbf{k}} \left(\Delta a_{\mathbf{k},\uparrow}^\dagger a_{-\mathbf{k},\downarrow}^\dagger + \Delta^* a_{-\mathbf{k},\downarrow} a_{\mathbf{k},\uparrow} \right) - \mathcal{V} \frac{|\Delta|^2}{U_{\text{eff}}}, \quad (8.63)$$

which is different from the quadratic mean-field Hamiltonian of a single-channel model. In the place of U, we have an effective interaction U_{eff} in the two-channel effective Hamiltonian. Obviously, we may follow the previous derivation and derive the gap equation in the two-channel model

$$-\frac{1}{U_{\text{eff}}} = \sum_{\mathbf{k}} \left(\frac{1}{2E_{\mathbf{k}}} - \frac{1}{2\epsilon_{\mathbf{k}}} \right). \quad (8.64)$$

For the number equation, we need to consider the population in both the closed-channel and open-channel. For the closed-channel population, we take the expectation value on both sides of Eq. (8.50) and notice that for the ground state at equilibrium, $\langle \dot{b}_0 \rangle = 0$, which leads to

$$\langle b_0 \rangle = -\frac{g}{\frac{\epsilon_{\mathbf{q}}}{2} + \gamma - 2\mu} \sum_{\mathbf{k}} \langle a_{\frac{\mathbf{q}}{2}-\mathbf{k},\downarrow} a_{\frac{\mathbf{q}}{2}+\mathbf{k},\uparrow} \rangle. \quad (8.65)$$

Combining this relation with Eq. (8.61), we find that the closed-channel population is related to the pairing gap

$$n_b = \left\langle b_0^\dagger b_0 \right\rangle = \frac{|\Delta|^2}{\left(g - \frac{U(\gamma - 2\mu)}{g} \right)^2} = \frac{|\Delta|^2}{\left(g_p - \frac{U_p(\gamma_p - 2\mu)}{g_p} \right)^2}. \quad (8.66)$$

The zero-temperature number equation is therefore

$$n = 2\frac{|\Delta|^2}{\left(g_p - \frac{U_p(\gamma_p - 2\mu)}{g_p} \right)^2} + \frac{1}{\mathcal{V}} \sum_{\mathbf{k}} \left(1 - \frac{\epsilon_{\mathbf{k}} - \mu}{E_{\mathbf{k}}} \right). \quad (8.67)$$

The chemical potential μ and the order parameter Δ can be solved self-consistently from the gap and number equations, which can then be used to calculate other physical parameters.

The two-channel model should be used whenever the closed-channel population is considerably large and would yield similar results as that of a single-channel model when the closed-channel population becomes negligible. Whether the closed-channel population is large or not depends on the nature of the Feshbach resonance under study. For the 834G resonance of ^{6}Li and the 202G resonance of ^{40}K, the closed-channel population is small. These resonances are called wide Feshbach resonances, characterized by the condition $g_p\sqrt{n}/E_F \gg 1$. For systems near these resonance points, the closed-channel population is negligibly small and it is sufficient to employ the single-channel model if one is only concerned with the physics near the unitary region.

As a concrete example, let us examine the case for ^{6}Li at 834G. The parameters for the ^{6}Li atoms are $a_{bg} = -1405a_0$, $\Delta B = -300G$ and $\mu_{co} \sim 2\mu_B$, where a_0 is the Bohr radius and μ_B is the Bohr magnetic moment [9]. In Fig. 8.4, we show the pairing gap and chemical potential throughout the crossover for ^{6}Li near the 834G resonance for both the single- and the two channel-models. In the strongly interacting region and on the BCS side, where the closed-channel population is small, the results from the two models are almost identical. Note that for the calculations in Fig. 8.4, we have simplified the problem by separating the 834G resonance in ^{6}Li from the other resonances. In reality, as there are plenty of scattering channels, there will be more than one Feshbach resonance for a given atom species. For example, in ^{6}Li, besides the s-wave Feshbach resonance

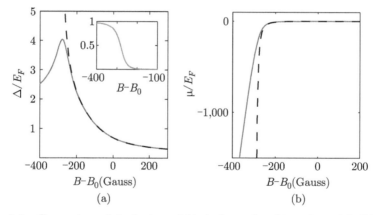

Figure 8.4.   Comparison of single-channel (dashed curve) and two-channel (solid curve) models for parameters of the $B_0 = 834G$ resonance of ^{6}Li. The unit of energy E_F is taken to be $E_F = \hbar^2(3\pi^2 n)^{2/3}/2m$, where the total particle density n is chosen to be a typical value in experiment $n \sim 2.9 \times 10^{13} \mathrm{cm}^{-3}$. The inset in (a) shows the closed-channel fraction.

at 834G between the low-lying hyperfine states $|F, m_F\rangle = |1/2, 1/2\rangle$ and $|F, m_F\rangle = |1/2, -1/2\rangle$, another s-wave Feshbach resonance between these states exists at 534 G [21]. Furthermore, there are also p-wave resonances in the system [22].

Another important situation in which one must apply the two-channel model is the narrow Feshbach resonance. For such resonances, the closed-channel population is typically large, which makes the single-channel model insufficient. We will introduce this type of Feshbach resonance in the next section.

8.5. Narrow Feshbach resonance

In the previous discussions, we have been modeling the two-body interactions as structureless contact interactions. Physically, this corresponds to the limit $r_0 k_F \ll 1$, where the effective interaction range r_0 characterizes the low-energy momentum dependence of the two-body scattering amplitude. In this limit, the energy scale of the short range interaction is much larger than all other relevant energy scales of the problem, and therefore may be modeled by a contact potential. In the opposite limit, i.e., $r_0 k_F \gg 1$, such a simplification fails naturally and a single-channel model with contact interaction can no longer characterize the system, as it does not provide the energy scale related to r_0. Feshbach resonances corresponding to the first case are called wide resonances and those corresponding to the latter, narrow resonances. Experimentally, whether a given resonance belongs to the wide or narrow Feshbach resonance depends on detailed scattering properties of the hyperfine states at a given magnetic field strength. More importantly, the most discussed resonances in the literature, i.e., 834G for ^{6}Li and 202G for ^{40}K, are both wide Feshbach resonances and one can describe the physics close to the unitary region using a single-channel model.

From the two-body scattering theory, we already know that the s-wave scattering amplitude can be written as

$$f(k) = \frac{e^{i2\delta} - 1}{ik}$$

$$= \frac{1}{k \cot \delta - ik}, \tag{8.68}$$

where the phase shift $\delta \propto k$ at low energies. Expanding the denominator to second-order in k, one gets

$$f(k) \approx -\frac{1}{-\frac{1}{2}r_0 k^2 + a^{-1} + ik}, \tag{8.69}$$

where r_0 and a^{-1} are expansion coefficients of $k \cot \delta$. In the low-energy limit, $f(0) \sim -a$, we recover the s-wave scattering length. In the intermediate energy range, however, the effects of the quadratic term involving the effective range r_0 cannot be neglected.

Microscopically, the narrow Feshbach resonance corresponds to a negative and large r_0 such that $|r_0|k_F \gg 1$. In this case, the two-body interaction potential at short range supports a long-lived quasi-bound state with positive energy that may decay into the continuum through an energy barrier (c.f. Ref. [15]). During the scattering process, the atoms spend a long time in the short-range quasi-bound state, which is in contrast to a wide Feshbach resonance.

For narrow Feshbach resonances, it has been shown that a two-channel model gives quantitatively accurate results throughout the crossover region [15]. From a two-body calculation with a two-channel model, one may establish the connection between the scattering parameters with the coupling coefficients in the Hamiltonian. More importantly, the effective range r_0 is related to the coupling rate between open and closed channels. This has two-fold implications. First, we may define a dimensionless parameter using the coupling rate g that differentiates wide and narrow Feshbach resonances

$$\gamma_w = \frac{g\sqrt{n}}{E_F} \sim \frac{1}{\sqrt{r_0 k_F}}. \tag{8.70}$$

Wide resonance corresponds to $\gamma_w \gg 1$ and narrow resonance corresponds to $\gamma_w \ll 1$. Second, for a narrow Feshbach resonance with $\gamma_w \ll 1$, it is possible to develop a perturbative theory based on the small parameter γ_w, even when the scattering length a_s diverges at the resonance. This is in contrast to a wide resonance, where there are no small parameters as a_s diverges.

From the analysis of the gap and number equations (see Ref. [15]), we see that the crossover region is essentially between $0 < \gamma_p/E_F < 2$. When the detuning γ_p is larger than $2E_F$ and almost all the atoms are in the open channel, the system is in the BCS limit and can be understood as a condensate of Cooper pairs at zero temperature. When the detuning is

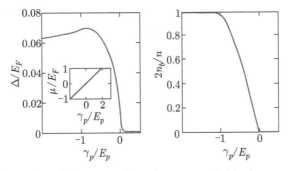

Figure 8.5. Illustration of the properties of a narrow Feshbach resonance. The unit of energy E_F is given by $E_F = \hbar^2 (3\pi^2 n)^{2/3}/2m$, where n is the total density of the gas. Note that the closed-channel population increases from close to 0 to close to 1 within a range of $\gamma_p/E_F \sim 1$, which corresponds to 0.027G for ^{6}Li with a typical density $2.9 \times 10^{13} \mathrm{cm}^{-3}$.

in the range of $0 < \gamma_p/E_F < 2$, the atoms are partially converted into molecules and the system is in a coherent mixture of pairing superfluidity and molecular condensation of molecules. The chemical potential goes through zero as γ_p becomes negative and all the atoms are converted into closed-channel molecules. This picture is illustrated in Fig. 8.5. Compared to a wide Feshbach resonance, the closed-channel population increases from 0 to the maximum possible value over a very narrow region and the system immediately becomes bosonic as the detuning becomes negative.

In the following, we will focus on wide Feshbach resonances, where a single-channel model is sufficient near the Feshbach resonance point.

8.6. BCS-BEC crossover in a harmonic trapping potential

In the previous sections, we have been discussing the properties of the BCS-BEC crossover for a uniform Fermi gas. Experimentally, to hold ultracold atoms in space, a trapping potential of some form is always present. Therefore, it is often useful to study the behavior of the Fermi gas near a Feshbach resonance in a harmonic trapping potential. There are two common practices: the local-density approximation (LDA) and the Bogoliubov–de-Gennes (BdG) equations [16]. We will discuss these two approaches in this section.

The LDA approach assumes that the trapping potential varies slowly enough so that each spatial point in the trap can be viewed as an infinitely large uniform system. This is essentially equivalent to the Thomas–Fermi approximation, where the local density is determined from the local Fermi energy. Under LDA, the local chemical potential $\mu(\mathbf{r})$ is related to that at the center of the trap μ as $\mu(\mathbf{r}) = \mu - V(\mathbf{r})$, where $V(\mathbf{r})$ is the trapping potential. The local order parameter $\Delta(\mathbf{r})$ can then be calculated from the gap equation using local chemical potentials as parameters. Finally, the chemical potential at the center of the trap can be determined self-consistently by the total particle number via the integrated number equation

$$N = \int d^3r\, n(\mathbf{r})$$

$$= \int d^3\mathbf{r} \left(1 - \frac{\epsilon_{\mathbf{k}} - \mu(\mathbf{r})}{E_{\mathbf{k}}(\mathbf{r})}\right), \qquad (8.71)$$

where $E_{\mathbf{k}} = \sqrt{(\epsilon_{\mathbf{k}} - \mu(\mathbf{r}))^2 + |\Delta(\mathbf{r})|^2}$.

In Fig. 8.6, we show the density distribution in a typical harmonic trap $V(\mathbf{r}) = \sum_{i=x,y,z} \frac{1}{2} m \omega_i^2 r_i^2$ using a single-channel model. We choose the units of energy to be the Fermi energy E_F at the trap center for N non-interacting trapped fermions with equal spin population for the two spin components. Correspondingly, we take the units of length along the i-th direction to be $R_i = \sqrt{2E_F/m\omega_i^2}$. The harmonic potential in dimensionless form becomes isotropic: $V(\mathbf{r})/E_F = \sum_i \tilde{\mathbf{r}}_i^2 \equiv \tilde{\mathbf{r}}^2$, with $\tilde{\mathbf{r}} = \mathbf{r}_i/R_i$. It is then straightforward to show the relation between E_F and the total

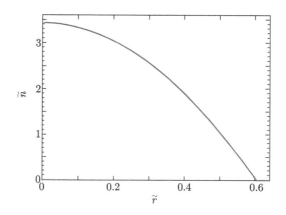

Figure 8.6.   Typical density distribution of a trapped Fermi gas under LDA.

particle number

$$N = \int d^3 \mathbf{r} n(\mathbf{r})$$

$$= \int d^3 \mathbf{r} \frac{1}{3\pi^2} \left[\frac{2m}{\hbar^2} \left(E_F - V(\mathbf{r}) \right) \right]^{\frac{3}{2}}$$

$$= \frac{1}{3\pi^2} \frac{8E_F^3}{\hbar^3 \omega_x \omega_y \omega_z} \int d^3 \tilde{\mathbf{r}} \left(1 - \tilde{r}^2 \right)^{\frac{3}{2}}$$

$$= \frac{E_F^3}{3\hbar^3 \omega_x \omega_y \omega_z}, \tag{8.72}$$

which gives $E_F = (3N\omega_x \omega_y \omega_z)^{\frac{1}{3}}$. With these units, the dimensionless form of the gap and number equation becomes

$$-\frac{1}{16}(k_F a_s)^{-1} = \int d^3 \tilde{\mathbf{k}} \left(\frac{1}{2\tilde{E}_{\tilde{\mathbf{k}}}} - \frac{1}{2\tilde{k}^2} \right), \tag{8.73}$$

$$1 = \frac{3}{\pi^3} \int d^3 \tilde{\mathbf{r}} \int d^3 \tilde{\mathbf{k}} \left(1 - \frac{\tilde{k}^2 - \tilde{\mu}(\tilde{\mathbf{r}})}{\tilde{E}_{\tilde{\mathbf{k}}}} \right), \tag{8.74}$$

where $\tilde{E}_{\tilde{\mathbf{k}}} = \sqrt{(\tilde{k}^2 - \tilde{\mu}(\tilde{\mathbf{r}}))^2 + |\tilde{\Delta}|^2}$. Apparently, the properties of the system only depend on the dimensionless variable $(k_F a_s)^{-1}$, which indicates the position in the BCS-BEC crossover. From the construction of our formalism, we see that the presence of an anisotropic harmonic potential with differing atom numbers do not add extra complexity to the problem. This nice feature comes from the LDA and the harmonicity of the trapping potential.

The LDA typically works very well toward the center of the trap, where the density of the gas is large and the curvature of the trapping potential is small. It fails near the edge of the trap, where the density distribution calculated from the LDA typically features a discontinuity in the spatial derivatives. Another important circumstance where the LDA fails is near the first-order phase boundaries, e.g., in a polarized Fermi gas. The order parameter and hence the density distribution calculated from the LDA typically shows an artificial discontinuity at the first-order phase boundaries. Realistically, the densities vary smoothly throughout. A way to circumvent this artefact is to go beyond the local-density approximation, for which a common practice is to use the BdG equations.

In principle, while the LDA is not limited to mean-field level calculations, the BdG equation is essentially a mean-field theory that takes the trapping potential into account. To derive the equations, we start from the second-quantized Hamiltonian in real space

$$
H - \mu N = \sum_{\sigma=\uparrow,\downarrow} \int d^3 \mathbf{r} \Psi_\sigma^\dagger(\mathbf{r}) \left(-\frac{\hbar^2 \nabla^2}{2m} - \mu + V(\mathbf{r}) \right) \Psi_\sigma(\mathbf{r})
$$

$$
- \int d^3 \mathbf{r} \left[\Delta(\mathbf{r}) \Psi_\uparrow^\dagger(\mathbf{r}) \Psi_\downarrow^\dagger(\mathbf{r}) + H.C. \right] - \int d^3 \mathbf{r} \frac{|\Delta(\mathbf{r})|^2}{U}, \quad (8.75)
$$

where we have already taken the mean field $\Delta(\mathbf{r}) = -U \langle \Psi_\downarrow(\mathbf{r}) \Psi_\uparrow(\mathbf{r}) \rangle$ with the assumption of a contact interaction potential $U(\mathbf{r} - \mathbf{r}') = 4\pi\hbar^2 \delta(\mathbf{r} - \mathbf{r}')/m$. In the presence of an external trapping potential $V(\mathbf{r})$, the pairing mean-field becomes position dependent.

The effective Hamiltonian Eq. (8.75) is quadratic and therefore can be diagonalized. Similar to the homogeneous case, we introduce the Bogoliubov transformation with position-dependent coefficients

$$
\Psi_\uparrow(\mathbf{r}) = \sum_i \left[u_i(\mathbf{r}) \alpha_{i,\uparrow} + v_i^*(\mathbf{r}) \alpha_{i,\downarrow}^\dagger \right], \quad (8.76)
$$

$$
\Psi_\downarrow(\mathbf{r}) = \sum_i \left[u_i(\mathbf{r}) \alpha_{i,\downarrow} - v_i^*(\mathbf{r}) \alpha_{i,\uparrow}^\dagger \right]. \quad (8.77)
$$

The Bogoliubov quasi-particle operators $\alpha_{i,\sigma}$ should satisfy the Fermi–Dirac statistics and leave the transformed effective Hamiltonian diagonal. The former requirement leads to the orthogonality relation of the coefficients:

$$
\int d^3 \mathbf{r} \left[u_i^*(\mathbf{r}) u_j(\mathbf{r}) + v_i^*(\mathbf{r}) v_j(\mathbf{r}) \right] = \delta_{ij}. \quad (8.78)
$$

Substituting the Bogoliubov transformation Eqs. (8.76) and (8.77) into the effective Hamiltonian Eq. (8.75), we get

$$
H - \mu N = \int d^3 \mathbf{r} \sum_{ij} \left(\alpha_{i,\downarrow} \; \alpha_{i,\uparrow}^\dagger \right) \begin{pmatrix} v_i(\mathbf{r}) & u_i(\mathbf{r}) \\ u_i^*(\mathbf{r}) & -v_i^*(\mathbf{r}) \end{pmatrix} \begin{pmatrix} H_0 & -\Delta(\mathbf{r}) \\ -\Delta^*(\mathbf{r}) & -H_0 \end{pmatrix}
$$

$$
\times \begin{pmatrix} v_j^*(\mathbf{r}) & u_j(\mathbf{r}) \\ u_j^*(\mathbf{r}) & -v_j(\mathbf{r}) \end{pmatrix} \begin{pmatrix} \alpha_{j,\downarrow}^\dagger \\ \alpha_{j,\uparrow} \end{pmatrix} + \left(H_0 - \int d^3 \mathbf{r} \frac{|\Delta|^2}{U} \right), \quad (8.79)
$$

with $H_0 = -\frac{\hbar^2 \nabla^2}{2m} - \mu + V(\mathbf{r})$. Under the requirement that the Hamiltonian be diagonal in the quasi-particle basis and making use of the orthogonality relation Eq. (8.78), we have

$$\begin{pmatrix} H_0 & \Delta(\mathbf{r}) \\ \Delta^*(\mathbf{r}) & -H_0 \end{pmatrix} \begin{pmatrix} u_i(\mathbf{r}) \\ v_i(\mathbf{r}) \end{pmatrix} = E_i \begin{pmatrix} u_i(\mathbf{r}) \\ v_i(\mathbf{r}) \end{pmatrix}. \tag{8.80}$$

This is the famous BdG equation and it is clear now that the index i indicates the i-th eigenstate, with E_i the i-th eigenenergy. Note that $\Delta(\mathbf{r})$ can be expressed in terms of the coefficients $u_i(\mathbf{r})$ and $v_i(\mathbf{r})$ by definition

$$\Delta(\mathbf{r}) = -U \langle \Psi_\downarrow(\mathbf{r}) \Psi_\uparrow(\mathbf{r}) \rangle$$

$$= -U \left\langle \sum_{ij} \left(u_i(\mathbf{r})\alpha_{i,\downarrow} - v_i^*(\mathbf{r})\alpha_{i,\uparrow}^\dagger \right) \left(v_i^*(\mathbf{r})\alpha_{j,\downarrow}^\dagger + u_i(\mathbf{r})\alpha_{j,\uparrow} \right) \right\rangle$$

$$= -U \sum_i u_i(\mathbf{r}) v_i^*(\mathbf{r}). \tag{8.81}$$

For practical purposes, the BdG equations (8.80) are typically expressed in the eigenbasis of H_0. Assuming $f_n(\mathbf{r})$ is the n-th eigenfunction of H_0, we take the following expansion of the coefficients in the BdG equations

$$u_i(\mathbf{r}) = \sum_n f_n(\mathbf{r}) u_n, \tag{8.82}$$

$$v_i(\mathbf{r}) = \sum_n f_n(\mathbf{r}) v_n. \tag{8.83}$$

We may then substitute these expressions into Eq. (8.80), multiply both sides with $f_m^*(\mathbf{r})$, integrate over $\mathbf{r}$ and apply the orthogonality relation of the eigenfunctions to get

$$\sum_n (H_{mn} u_n + \Delta_{mn} v_n) = \epsilon_i u_m, \tag{8.84}$$

$$\sum_n (\Delta_{mn}^* u_n - H_{mn} v_n) = \epsilon_i v_m, \tag{8.85}$$

where $H_{mn} = \int d^3\mathbf{r} f_m^*(\mathbf{r}) H_0 f_n(\mathbf{r})$ and $\Delta_{mn} = \int d^3\mathbf{r} f_m^*(\mathbf{r}) \Delta(\mathbf{r}) f_n(\mathbf{r})$. In principle, Eqs. (8.81) and (8.84) can now be solved numerically. For a detailed review, the readers may refer to Ref. [17]

An important application of the BdG equation in these Fermi gases is the ability to characterize vortices of the pairing superfluidity, just as the Gross–Pitaeveskii (GP) equations were used to characterize vortex structures in a BEC [18, 19]. As vortices are typically associated with rotations

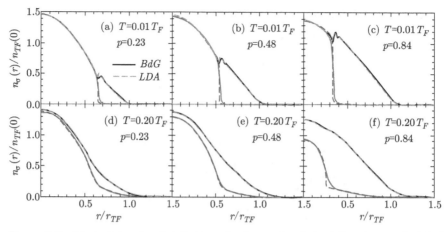

Figure 8.7. Typical density distribution of a trapped polarized Fermi gas calculated using BdG equations. From Ref. [17].

around a given axis, the simplest numeric scenario is a rotating gas in a harmonic trapping potential with cylindrical symmetry. In this case, the order parameter can be assumed to take the form $\Delta(\mathbf{r}) = \Delta(\rho)e^{-i\theta}$ and the Bogoliubov quasi-particle wavefunctions can be expanded over the eigenfunctions of a cylindrical harmonic potential [18]

$$u_n(\mathbf{r}) = \sum_j c_{nj}\phi_{jl}(\rho)e^{il\theta}e^{ik_z z}, \qquad (8.86)$$

$$v_n(\mathbf{r}) = \sum_j d_{nj}\phi_{jl+1}(\rho)e^{i(l+1)\theta}e^{ik_z z}, \qquad (8.87)$$

with $\phi_{jl}(\rho) = \sqrt{2}J_l(\alpha_{jl}\rho/R)/[RJ_{l+1}(\alpha_{jl})]$ and α_{jl} the j-th root of the spherical Bessel function $J_l(x)$. Similar to the previous case without angular momentum, one can show that different subspaces of l and k_z are only coupled through the gap and number equations.

Note that in the absence of an external trapping potential, the eigenstate of H_0 consists of plane waves. Considering that the coefficients u_i and v_i are now homogeneous in space, we may replace u_i with $e^{i\mathbf{k}\cdot\mathbf{r}}u_{\mathbf{k}}$ and v_i with $e^{i\mathbf{k}\cdot\mathbf{r}}v_{\mathbf{k}}$, so that Eq. (8.80) reduces to

$$\begin{pmatrix} H_0 & \Delta \\ \Delta^* & -H_0 \end{pmatrix} \begin{pmatrix} u_{\mathbf{k}} \\ v_{\mathbf{k}} \end{pmatrix} = \begin{pmatrix} u_{\mathbf{k}} \\ v_{\mathbf{k}} \end{pmatrix}, \qquad (8.88)$$

which reproduces the results we previously got for a uniform gas.

Finally, in the strongly interacting region, it is possible to manipulate the BdG equations into the form of a GP equation for the BEC [20]. As the BdG equations are essentially on the mean-field level, the coefficient in the GP equation suggests a dimer-dimer scattering length of $a_d = 2a_s$, just as one would expect.

References

[1] G. D. Mahan. *Many-particle Physics*, 3rd Ed. Springer, Berlin (2000).

[2] L. N. Cooper. *Phys. Rev.* **104**, 1189 (1956).

[3] M. P. Marder. *Condensed Matter Physics*. John Wiley & Sons, New York (2000).

[4] J. Bardeen, L. N. Cooper and J. R. Schrieffer. *Phys. Rev.* **108**, 1175 (1957).

[5] N. N. Bogoliubov. *JETP* **7**, 41 (1958).

[6] D. M. Eagles. *Phys. Rev.* **186**, 456 (1969).

[7] A. J. Leggett. *Modern Trends in the Theory of Condensed Matter*. Springer-Verlag, Berlin (1980).

[8] M. Marini, F. Pistolesi and G. C. Strinati. *Eur. Phys. J. B* **1**, 151 (1998).

[9] W. Ketterle and M. W. Zwierlein. *Proceedings of the International School of Physics "Enrico Fermi"*, Course CLXIV. Varenna (2006).

[10] D. S. Petrov, C. Salomon and G. V. Shlyapnikov. *Phys. Rev. Lett.* **93**, 090404 (2004).

[11] J. Carlson, S.-Y. Chang, V. R. Pandharipande and K. E. Schmidt. *Phys. Rev. Lett.* **91**, 050401 (2003).

[12] M. Holland, S. J. J. M. F. Kokkelmans, M. L. Chiofalo and R. Walser. *Phys. Rev. Lett.* **87**, 120406 (2001).

[13] J. Levinsen and V. Gurarie. *Phys. Rev. A* **73**, 053607 (2006).

[14] Q. Chen, J. Stajic, S. Tan and K. Levin. *Phys. Rep.* **412**, 1 (2005).

[15] D. E Sheehy and L. Radzihovsky. *Ann. Phys.* **322**, 1790 (2007).

[16] P.-G. de Gennes. *Superconductivity of Metals and Alloys*. Benjamin, New York (1966).

[17] X.-J. Liu, H. Hu and P. D. Drummond. *Phys. Rev. A* **75**, 023614 (2007).

[18] R. Sensarma, M. Randeria and T. L. Ho. *Phys. Rev. Lett.* **96**, 090403 (2006).

[19] C.-C. Chien, Y. He, Q. Chen and K. Levin. *Phys. Rev. A* **73**, 041603(R) (2006).

[20] P. Pieri and G. C. Strinati. *Phys. Rev. Lett.* **91**, 030401 (2003).

[21] K. Dieckmann, C. A. Stan, S. Gupta, Z. Hadzibabic, C. H. Schunck and W. Ketterle. *Phys. Rev. Lett.* **89**, 203201 (2002).

[22] C. H. Schunck, M. W. Zwierlein, C. A. Stan, S. M. F. Raupach, W. Ketterle, A. Simoni, E. Tiesinga, C. J. Williams and P. S. Julienne. *Phys. Rev. A* **71**, 045601 (2005).

9
Beyond-Mean-Field Descriptions

So far, our focus has been on the mean-field description of the BCS-BEC crossover, which can be regarded as an extension of the standard BCS theory. While in three dimensions, the BCS wave function provides a qualitatively correct picture for the BCS-BEC crossover at zero temperature and the mean-field theory we reviewed previously cannot be expected to yield quantitatively correct results in a strongly interacting system. On the other hand, the success of the standard BCS theory is largely due to the lack of considerable quantum fluctuations in the weak-interacting BCS limit, which is the case for the conventional superconducting materials. When the interaction strength is increased toward unitarity, the quantum fluctuation increases and must be taken into account at some point. At finite temperatures, one must also include the thermal fluctuations and the problem becomes more complicated. In the strongly interacting regime $(-1 < (k_F a_s)^{-1} < 1)$, the effective s-wave scattering length becomes very large and diverges at unitarity. Because of the lack of small parameters, the system cannot be characterized by a reliable perturbative theory. One therefore has to either resort to exact solutions via, for example, Quantum Monte Carlo methods, or rely on truncating fluctuation expansions based on physical intuitions.

In this chapter, we discuss several beyond-mean-field theories of a BCS-BEC crossover. We will see that while they all provide more accurate characterization of the physical parameters near the unitary region than the mean-field theory does, there still exist subtle differences between these theories. As this chapter is not intended as an exhaustive review of the many-body BCS-BEC crossover schemes, we limit ourselves to a modest outline of the typical approaches. We will point interested readers to the relevant references for more details.

9.1.   NSR scheme

In their seminal 1985 paper, Nozieres and Schmitt-Rink worked out the critical temperature T_c for the onset of BCS superfluidity, by including the Gaussian fluctuations near the critical point [1]. This is achieved by expanding the effective action of the system to the quadratic order at the critical point when the pairing order parameter vanishes. By taking thermal fluctuations into account, the so-called NSR scheme produces a much lower transition temperature at unitarity than the mean-field theory. It paves the way for the development of more sophisticated many-body calculations. In this section, we will briefly outline the NSR scheme.

Let us first examine the fluctuations in the thermodynamic potential of a repulsively interacting two-component Fermi gas, where the Cooper instability is absent [2, 3]. Assuming contact s-wave scattering potential $U(\mathbf{r}) = U\delta(\mathbf{r})$ between fermions of different spin species, we can describe the two-body scattering process via the Lippmann–Schwinger equation

$$
\begin{aligned}
T(Q) &= U - U \sum_K G_0(K)G_0(Q-K)T(Q) \\
&= U - U\chi(Q)T \\
&= \frac{U}{1 + U\chi(Q)},
\end{aligned}
$$
(9.1)

where $G_0 = (E - H_0)^{-1}$ is the Green's function for a free fermion. Note for brevity, we follow the four-vectors notation [4]: $K = (\mathbf{k}, i\omega_n)$, $Q = (\mathbf{q}, i\Omega_n)$ and $\sum_K = \frac{1}{\beta}\sum_{\mathbf{k}, i\omega_n}$. Here, ω_n (Ω_n) is the Matsubara frequency for fermions (bosons). In Eq. (9.1), we have further suppressed the summation on the second line. The pair susceptibility $\chi(Q) = \frac{1}{\beta}\sum_K G_0(-K)G_0(K + Q)$, with the four-vectors $K = (\mathbf{k}, i\omega_n)$. The bare interaction rate U should be renormalized in the standard fashion: $1/U = m/(4\pi\hbar^2 a_s) - \sum_{\mathbf{k}} 1/(2\epsilon_{\mathbf{k}})$. Equation (9.1) corresponds to the ladder diagram illustrated in Fig. 9.1(a). The full Green's function of a single fermion is dressed by these scattering processes through the Dyson's equation

$$\Sigma(K) = \frac{1}{\beta} \sum_Q T(Q) G_0(Q - K), \tag{9.2}$$

$$G(K) = G_0(K) + G_0(K)\Sigma(K)G(K)$$
$$\approx G_0(K) + G_0(K)\Sigma(K)G_0(K), \tag{9.3}$$

where only the leading order in the Dyson's expansion is retained. The diagrammatical schemes for these relations are shown in Fig. 9.1.

Naively, one may try and extend the formalism above to an attractively interacting Fermi gas. However, due to the existence of the Cooper instability below the critical temperature, a pole develops in the T-matrix in Eq. (9.1), rendering the calculation invalid. On the other hand, for a thermal Fermi gas with attractive interaction ($T > T_c$), nothing should prevent us from characterizing the system using Eqs. (9.1) to (9.3) by simply changing the sign of the scattering length a_s in the expression for the bare interaction rate U. Therefore, it is reasonable to assume that the description for a repulsively interacting Fermi gas works for the attractively interacting case

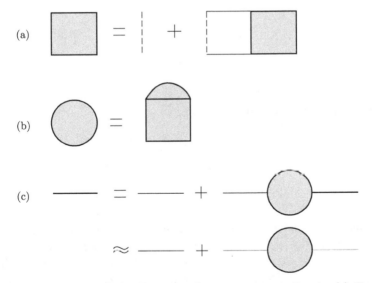

Figure 9.1.   Illustration of the Feynman diagrams corresponding to (a) Eq. (9.1), (b) Eq. (9.2) and (c) Eq. (9.3). The thin solid lines represent G_0, the thick solid lines represent the full Green's function G and the dashed lines represent interaction U.

down to the critical temperature. The critical temperature is then characterized by the onset of the pole in the corresponding T-matrix. Indeed, it
has been shown in Ref. [3] that the formalism for a repulsive gas is the same
as that of a superfluid state in the $\Delta \to 0$ limit, provided that the sign of
the s-wave scattering length is changed. These considerations lead to the
Thouless condition for the critical temperature

$$1 + U\chi(0) = 0, \tag{9.4}$$

where we have assumed that the pole in the T-matrix first develops in the
$Q = 0$ sector.

To gain further insight of the Thouless condition, we may carry out the
Mastubara sum explicitly

$$
\begin{aligned}
\chi(0) &= \frac{1}{\beta} \sum_{\mathbf{k},\omega_n} G_0(-\mathbf{k}, -i\omega_n) G_0(\mathbf{k}, i\omega_n) \\
&= \frac{1}{\beta} \sum_{\mathbf{k},\omega_n} \frac{1}{-i\omega_n - \xi_{\mathbf{k}}} \frac{1}{i\omega_n - \xi_{\mathbf{k}}} \\
&= -\frac{1}{\beta} \sum_{\mathbf{k},\omega_n} \frac{1}{2\xi_{\mathbf{k}}} \left[\frac{1}{i\omega_n - \xi_{\mathbf{k}}} - \frac{1}{i\omega_n + \xi_{\mathbf{k}}} \right] \\
&= \sum_{\mathbf{k}} \frac{1}{2\xi_{\mathbf{k}}} \tanh \frac{\beta\xi_{\mathbf{k}}}{2}.
\end{aligned}
\tag{9.5}
$$

Equation (9.4) then becomes

$$-\frac{1}{U} = \sum_{\mathbf{k}} \frac{1}{2\xi_{\mathbf{k}}} \tanh \frac{\beta\xi_{\mathbf{k}}}{2}, \tag{9.6}$$

which is exactly the finite-temperature mean-field gap equation at $\Delta = 0$.

Of course, the chemical potential is still unknown and needs to be
determined from the number equation, similar to the mean-field BCS-BEC
crossover theory. It is here where the main difference between the NSR
scheme and the mean-field calculations lies. Physically, for an NSR calculation, we need to include the contribution from the thermal fluctuations
in a modified number equation. Indeed, we will see that the fluctuations
modify the thermodynamic potential Ω and give rise to an additional term
while evaluating $n = -\partial\Omega/\partial\mu$.

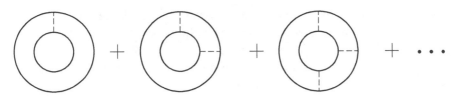

Figure 9.2.   Illustration of the Feynman diagrams that correspond to the thermodynamic potential Ω_f in Eq. (9.7). The solid lines represent G_0 and the dashed lines represent the bare interaction U.

The NSR scheme takes into account the fluctuations illustrated by the diagram in Fig. 9.1. This is consistent with the truncation in the Dyson's expansion in Eq. (9.3). The contribution to the thermodynamic potential by these fluctuations can be summed up as

$$\Omega_f = \Omega - \Omega_0 = \frac{2}{\beta} \sum_{l=1}^{\infty} (-1)^l \frac{U^l}{l} \sum_Q \chi^l(Q)$$

$$= \frac{2}{\beta} \sum_Q \ln\left[1 - U\chi(Q)\right], \qquad (9.7)$$

where Ω_0 represents the thermodynamic potential at the mean-field level.

And the modified number equation becomes

$$n = -\frac{\partial \Omega_0}{\partial \mu} - \frac{\partial \Omega_f}{\partial \mu}. \qquad (9.8)$$

The critical temperature can be evaluated by solving Eqs. (9.6) and (9.8) together. For details, interested readers may refer to Refs. [1] and [3].

Notice that although the number equation is modified using the thermal fluctuations, the gap equation is left unchanged. The NSR scheme is therefore not self-consistent. However, maintaining the form of the gap equation is essential to ensure that the low-energy excitation of the gas is gapless, a physical consequence of the broken symmetry phase below T_c [5].

9.2.   Path integral and saddle point expansion

In 1993, Sà de Melo *et al.* extended the NSR scheme to describe the broken symmetry phase below the critical temperature [6]. In contrast to the NSR calculations, these authors included the Gaussian fluctuations near the saddle point of the effective action, which corresponded to the stationary phase with a non-vanishing order parameter Δ. Later, various authors have developed schemes for the broken symmetry phase based on similar expansions, but with different choices of the gap and number equations. In this section, we will outline these many-body approaches. For a natural presentation, we will discuss the formalism in the language of path integrals.

At a finite temperature T, the grand canonical partition function Z can be written in the form of path integrals [7]

$$Z = \text{Tr} e^{-\beta[H-\mu N]}$$

$$= \int \mathcal{D}\left[\psi^\dagger, \psi\right] e^{-S[\psi^\dagger, \psi]}, \qquad (9.9)$$

where the effective action is defined as

$$S\left[\psi^\dagger, \psi\right] = \int_0^\beta d\tau \left\{ \int d^3\mathbf{r} \psi^\dagger(x) \left(\frac{\partial}{\partial \tau} - \mu \right) \psi(x) + H \right\}, \qquad (9.10)$$

with $x = (\mathbf{r}, \tau)$. Here τ is the imaginary time. The effective single-channel Hamiltonian H in Eq. (9.10) is

$$H = \int dx \psi_\sigma^\dagger(x) \left(-\frac{\hbar^2 \nabla^2}{2m} \right) \psi_\sigma(x) + U \int dx \psi_\uparrow^\dagger(x) \psi_\downarrow^\dagger(x) \psi_\downarrow(x) \psi_\uparrow(x). \qquad (9.11)$$

Again, we have assumed a contact interaction potential, which needs to be renormalized later following the standard procedure.

To proceed, one must bring the Hamiltonian and hence the effective action into the Gaussian form. This can be achieved via the Hubbard–Stratonovich transformation, for which an auxiliary bosonic field η is introduced

$$\eta(x) = \psi_\downarrow(x) \psi_\uparrow(x). \qquad (9.12)$$

The interaction term in the effective action then becomes quadratic

$$U \int dx \psi_\uparrow^\dagger(x) \psi_\downarrow^\dagger(x) \psi_\downarrow(x) \psi_\uparrow(x) = \int dx \eta^\dagger(x) U \eta(x). \qquad (9.13)$$

Once in the Gaussian form, we may integrate out the bosonic degrees of freedom via the following identity [7]

$$[\Delta H]^{-1} e^{\sum_{\alpha,\beta} \eta_\alpha^* H_{\alpha\beta}^{-1} \eta_\beta} = \int \frac{1}{2\pi i} \prod_\alpha d\xi_\alpha^* d\xi_\alpha e^{-\sum_{\alpha,\beta} \xi_\alpha^* H_{\alpha\beta} \xi_\beta + \sum_\alpha (\eta^* \xi_\alpha + \eta_\alpha \xi_\alpha^*)},$$

$$(9.14)$$

where ξ_α, η_α are complex variables representing bosonic fields. We set $\xi = \Delta$ and $H_{\alpha\beta}^{-1} = -U$, such that the indices α and β drop out

$$e^{-\int dx \eta^\dagger(x) U \eta(x)} = \int \mathcal{D}[\Delta^*, \Delta] e^{\int dx \left\{ \frac{|\Delta|^2}{U} + \psi_\uparrow^\dagger(x) \psi_\downarrow^\dagger(x) \Delta(x) + \psi_\downarrow(x) \psi_\uparrow(x) \Delta^*(x) \right\}}.$$

$$(9.15)$$

The grand partition function can now be written as

$$Z = \int \mathcal{D}[\Delta^*, \Delta] \int \mathcal{D}[\psi^\dagger, \psi] e^{\int dx \frac{|\Delta|^2}{U}}$$

$$\times e^{-\int dx \left[\psi_\sigma^* \left(-\frac{\hbar^2 \nabla^2}{2m} + \partial_\tau - \mu \right) \psi_\sigma - \psi_\uparrow^\dagger \psi_\downarrow^\dagger \Delta - \psi_\downarrow \psi_\uparrow \Delta^* \right]}$$

$$= \int \mathcal{D}[\Delta^*, \Delta] \int \mathcal{D}[\psi^\dagger, \psi] e^{\int dx \frac{|\Delta|^2}{U} + \mu} e^{\int dx \zeta^\dagger(x) \mathcal{G}^{-1}(x) \zeta(x)}, \qquad (9.16)$$

where $\zeta(x) = \begin{pmatrix} \psi_\uparrow(x) \\ \psi_\downarrow^\dagger(x) \end{pmatrix}$ and the inverse Nambu propagator $\mathcal{G}^{-1}(x)$ is defined as

$$\mathcal{G}^{-1}(x) = \begin{pmatrix} -\partial_\tau + \frac{\hbar^2 \nabla^2}{2m} + \mu & \Delta(x) \\ \Delta^*(x) & -\partial_\tau - \frac{\hbar^2 \nabla^2}{2m} - \mu \end{pmatrix}. \qquad (9.17)$$

In deriving the last line of Eq. (9.16), we have used the property of $\psi_\sigma(x)$ as the Grassmann variables and integrated by part to get

$$\int dx \psi_\downarrow^\dagger(x) \left(-\frac{\hbar^2 \nabla^2}{2m} + \partial_\tau - \mu \right) \psi_\downarrow(x) = \int dx \psi_\downarrow(x) \left(\frac{\hbar^2 \nabla^2}{2m} + \partial_\tau + \mu \right) \psi_\downarrow^\dagger(x).$$

$$(9.18)$$

We may now integrate out the fermionic fields in the partition function using the Gaussian integral for Grassmann variables

$$[\det H] e^{\sum_{\alpha,\beta} \eta_\alpha^* H_{\alpha\beta}^{-1} \eta_\beta} = \int \prod_\alpha d\xi_\alpha^* d\xi_\alpha e^{-\sum_{\alpha,\beta} \xi_\alpha^* H_{\alpha\beta} \xi_\beta + \sum_\alpha (\eta^* \xi_\alpha + \eta_\alpha \xi_\alpha^*)}.$$

$$(9.19)$$

Setting $\eta = 0$, $H = \mathcal{G}^{-1}$ and $\xi = \zeta$, we have

$$Z = \int \mathcal{D}\left[\Delta^*, \Delta\right] e^{\int dx \left(\frac{|\Delta|^2}{U} + \mu\right)} \det \mathcal{G}^{-1}(x)$$

$$= \int \mathcal{D}\left[\Delta^*, \Delta\right] e^{\int dx \left(\frac{|\Delta|^2}{U} + \mu\right) + \ln\left(\det \mathcal{G}^{-1}\right)}$$

$$= \int \mathcal{D}\left[\Delta^*, \Delta\right] e^{\int dx \left(\frac{|\Delta|^2}{U} + \mu\right) + \mathrm{Tr}\left(\ln \mathcal{G}^{-1}\right)}$$

$$\equiv \int \mathcal{D}\left[\Delta^*, \Delta\right] e^{-S_{\mathrm{eff}}[\Delta^*, \Delta]}, \tag{9.20}$$

where we have used the equality $\ln\left(\det \mathcal{G}^{-1}\right) = \mathrm{Tr}\left(\ln \mathcal{G}^{-1}\right)$, with the trace running over the Nambu indices.

We wish to investigate fluctuations near the stationary point of the effective action. Therefore, following the stationary-phase approximation, we expand the bosonic field Δ around Δ_0,

$$\Delta(x) = \Delta_0 + \varphi(x), \tag{9.21}$$

where Δ_0 satisfies the stationary-phase condition $\partial S_{\mathrm{eff}}(\Delta_0, \Delta_0^*)/\partial \Delta_0^* = 0$. As we will see in the following, $\varphi(x)$ represents fluctuations of the superfluid order parameter above the mean-field solution Δ_0. In general, especially in the strong-coupling limit where the effects of fluctuations are large, there are no small parameters in the problem. Nevertheless, there are not many things that we can do other than expanding the effective action to a quadratic order in $\varphi(x)$. We expect that fluctuations of such a form can be calculated via a Gaussian integral.

Substituting Eq. (9.21) into the effective action S_{eff} in Eq. (9.20), we have

$$S_{\mathrm{eff}}\left[\varphi, \varphi^*\right] = \int dx \left(-\frac{|\Delta_0 + \varphi(x)|^2}{U} - \mu\right) - \mathrm{Tr}\left[\ln\left(\mathcal{G}_0^{-1} + V\right)\right], \tag{9.22}$$

where $\mathcal{G}_0^{-1} \equiv \mathcal{G}^{-1}\big|_{\Delta(x)=\Delta_0}$ and $V = \begin{pmatrix} 0 & \varphi \\ \varphi^* & 0 \end{pmatrix}$. We have also followed the common practice and denoted $\mathrm{Tr} = \int dx \mathrm{tr}$.

For the second term on the right-hand side of Eq. (9.22), we have the following expansion

$$
\mathrm{Tr}\left[\ln\left(\mathcal{G}_0^{-1} + V\right)\right]
$$
$$
= \mathrm{Tr}(\ln\mathcal{G}_0^{-1}) + \mathrm{Tr}[\ln\left(1 + \mathcal{G}_0 V\right)]
$$
$$
= \mathrm{Tr}(\ln\mathcal{G}_0^{-1}) + \mathrm{Tr}\left(\mathcal{G}_0 V\right) - \frac{1}{2}\mathrm{Tr}\left(\mathcal{G}_0 V \mathcal{G}_0 V\right) + O(\varphi^2). \qquad (9.23)
$$

The shorthands $\mathrm{Tr}\left(\mathcal{G}_0 V\right)$ and $\mathrm{Tr}\left(\mathcal{G}_0 V \mathcal{G}_0 V\right)$ respectively represent

$$
\mathrm{Tr}\left(\mathcal{G}_0 V\right) = \int dx_1 \mathrm{tr}\left[\mathcal{G}_0(x_1, x_1)V(x_1)\right],
$$
$$
\mathrm{Tr}\left(\mathcal{G}_0 V \mathcal{G}_0 V\right) = \int dx_1 dx_2 \mathrm{tr}\left[\mathcal{G}_0(x_1, x_2)V(x_2)\mathcal{G}_0(x_2, x_1)V(x_1)\right], \qquad (9.24)
$$

with $\mathcal{G}_0(x, x') = \mathcal{G}_0(x' - x)$ assuming translational symmetry.

Correspondingly, the effective action can be written as

$$
S_{\mathrm{eff}} = S_0 + S_1 + S_2 + O(\varphi^2)
$$
$$
= \int dx \left(-\frac{|\Delta_0|^2}{U} - \mu\right) - \mathrm{Tr}\ln\mathcal{G}_0^{-1} - \int dx \left(\frac{\Delta_0^*}{U}\varphi + \frac{\Delta_0}{U}\varphi^*\right) - \mathrm{Tr}\left(\mathcal{G}_0 V\right)
$$
$$
- \int dx \frac{|\varphi|^2}{U} + \mathrm{Tr}\left(\mathcal{G}_0 V \mathcal{G}_0 V\right) + O(\varphi^2). \qquad (9.25)
$$

We now analyze the expansion order-by-order.

The zeroth-order term S_0 corresponds to the effective action under the stationary-phase approximation, which leads to the mean-field results. For example, the mean-field gap equation can be derived from the stationary phase condition $\partial S_0 / \partial \Delta_0^* = 0$

$$
\frac{\partial S_0}{\partial \Delta_0^*} = \int dx \left[-\frac{\Delta_0}{U} - \mathrm{tr}\left(\frac{\partial}{\partial\Delta_0}\ln\mathcal{G}_0^{-1}\right)\right]
$$
$$
= -\beta\frac{\Delta_0}{U} - \int dx \mathrm{tr}\left(\mathcal{G}_0\frac{\partial}{\partial\Delta_0}\mathcal{G}_0^{-1}\right)
$$
$$
= -\beta\frac{\Delta_0}{U} - \sum_{\mathbf{k},\omega_n} \mathrm{tr}\left(\mathcal{G}_0(\mathbf{k}, i\omega_n)\frac{\partial}{\partial\Delta_0}\mathcal{G}_0^{-1}(\mathbf{k}, i\omega_n)\right) = 0, \qquad (9.26)
$$

where in the last step, we have Fourier-transformed the saddle-point Nambu propagator $\mathcal{G}_0$ to momentum-frequency space. The Nambu Green's function in momentum space can be written as [8]

$$\mathcal{G}_0^{-1}(\mathbf{k}, i\omega_n) = \begin{pmatrix} i\omega_n - \xi_\mathbf{k} & \Delta_0 \\ \Delta_0^* & i\omega_n + \xi_\mathbf{k} \end{pmatrix}, \tag{9.27}$$

$$\mathcal{G}_0(\mathbf{k}, i\omega_n) = \frac{1}{-\omega_n^2 - E_\mathbf{k}^2} \begin{pmatrix} i\omega_n + \xi_\mathbf{k} & -\Delta_0 \\ -\Delta_0^* & i\omega_n - \xi_\mathbf{k} \end{pmatrix}, \tag{9.28}$$

with the quasi-particle self-energy $E_\mathbf{k} = \sqrt{\xi_\mathbf{k}^2 + |\Delta_0|^2}$. It then follows that Eq. (9.26) leads to

$$-\frac{\beta}{U} = -\sum_{\mathbf{k}, \omega_n} \frac{1}{(i\omega_n + E_\mathbf{k})(i\omega_n - E_\mathbf{k})}. \tag{9.29}$$

Carrying out the Matsubara summation over the imaginary frequencies, we arrive at the mean-field gap equation

$$-\frac{1}{U} = \sum_\mathbf{k} \frac{1}{E_\mathbf{k}} \tanh \frac{\beta E_\mathbf{k}}{2}. \tag{9.30}$$

Similarly, the mean-field number equation can be derived using

$$N = -\frac{1}{\beta} \frac{\partial S_0}{\partial \mu}. \tag{9.31}$$

We now turn to the first-order expansion S_1. It is straightforward to show that this term vanishes at the stationary point $\Delta = \Delta_0$, as dictated by the mean-field gap equation Eq. (9.30). First of all we have

$$\begin{aligned} \mathrm{Tr}\,(\mathcal{G}_0 V) &= \int dx \mathrm{tr}\,[\mathcal{G}_0(x, x) V(x)] \\ &= \sum_{\mathbf{k}, \omega_n} \mathrm{tr}\,[\mathcal{G}_0(\mathbf{k}, i\omega_n) V(0)] \\ &= -\sum_{\mathbf{k}, \omega_n} \left(\frac{\Delta_0 \varphi^*}{-\omega_n^2 - E_\mathbf{k}} + \frac{\Delta_0^* \varphi}{-\omega_n^2 - E_\mathbf{k}} \right), \end{aligned} \tag{9.32}$$

where $\mathcal{G}_0(\mathbf{k}, i\omega_n)$ is given in Eq. (9.28) and $V(\mathbf{q}, i\nu_n) = \begin{pmatrix} 0 & \varphi(\mathbf{q}, i\nu_n) \\ \varphi(-\mathbf{q}, -i\nu_n) & 0 \end{pmatrix}$. Therefore,

$$S_1 = -\int dx \left(\frac{\Delta_0^*}{U} \varphi + \frac{\Delta_0}{U} \varphi^* \right) - \mathrm{Tr}\,(\mathcal{G}_0 V)$$

$$= -\sum_{\mathbf{k},\omega_n} \left(\frac{1}{U} - \frac{1}{-\omega_n^2 - E_{\mathbf{k}}} \right) \Delta_0 \varphi^*(0) + H.C.,$$

$$= 0. \tag{9.33}$$

Finally, the second-order expansion S_2 contains the Gaussian fluctuations near the stationary point. Similar to what we have done for the zeroth-order term, we need to Fourier transform the expression for S_2 to work out a simple expression in the desired quadratic form

$$S_2 = -\int dx \frac{|\varphi|^2}{U} + \frac{1}{2} \mathrm{Tr}\,(\mathcal{G}_0 V \mathcal{G}_0 V)$$

$$= \sum_{\mathbf{k},\omega_n} \left\{ -\frac{|\varphi|^2}{U} + \frac{1}{2} \sum_{\mathbf{q},\nu_n} \mathrm{tr}[G_0(\mathbf{k}, i\omega_n)V(\mathbf{q}, i\nu_n) \right.$$

$$\left. G_0(\mathbf{k} + \mathbf{q}, i\omega_n + i\nu_n)V(-\mathbf{q}, -i\nu_n)] \right\}, \tag{9.34}$$

As the effective action in Eq. (9.34) is quadratic in $\{\varphi, \varphi^*\}$, it is possible to re-arrange the expression into a Gaussian form, so that the partition function corresponding to the fluctuation $Z = \int \mathcal{D}\,[\varphi, \varphi^*]\,e^{-S_2}$ can be evaluated. From Eq. (9.34), one can show that

$$S_2 = \frac{1}{2} \sum_{\mathbf{q},\nu_n} \left(\varphi^*(\mathbf{q}, i\nu_n)\ \varphi(-\mathbf{q}, -i\nu_n) \right) M(\mathbf{q}, i\nu_n) \begin{pmatrix} \varphi(\mathbf{q}, i\nu_n) \\ \varphi^*(-\mathbf{q}, -i\nu_n) \end{pmatrix}, \tag{9.35}$$

where the matrix elements of the inverse fluctuation propagator $M(\mathbf{q})$ are

$$M_{11}(\mathbf{q}, i\nu_n) = M_{22}(-\mathbf{q}, -i\nu_n)$$

$$= -\frac{1}{U} + \sum_{\mathbf{k},\omega_n} \frac{i\omega_n + \xi_{\mathbf{k}}}{-\omega_n^2 - E_{\mathbf{k}}^2} \frac{i\omega_n + i\nu_n - \xi_{\mathbf{k+q}}}{(i\omega_n + i\nu_n)^2 - E_{\mathbf{k+q}}^2}, \tag{9.36}$$

$$M_{12}(\mathbf{q}, i\nu_n) = M_{21}(\mathbf{q}, i\nu_n) = \sum_{\mathbf{k},\omega_n} \frac{\Delta_0^2}{-\omega_n^2 - E_{\mathbf{k}}^2} \frac{1}{(i\omega_n + i\nu_n)^2 - E_{\mathbf{k+q}}^2}. \tag{9.37}$$

And the total partition function

$$Z = \int \mathcal{D}\left[\Delta, \Delta^*\right] e^{-S}$$

$$\approx e^{-S_0} \int \mathcal{D}\left[\varphi, \varphi^*\right] e^{-S_2}$$

$$= e^{-S_0} \sum_{\mathbf{q}, \nu_n} \left[\det M(\mathbf{q}, i\nu_n)\right]^{-\frac{1}{2}}, \tag{9.38}$$

where we have used the formula [7]

$$\int \mathcal{D}\left(\psi^*, \psi\right) e^{-\frac{1}{2}\left(\psi^* \ \psi\right) D \begin{pmatrix} \psi \\ \psi^* \end{pmatrix}} = \left(\det D\right)^{-\frac{1}{2}}. \tag{9.39}$$

Correspondingly, the thermodynamic potential

$$\Omega = -\frac{1}{\beta} \ln Z,$$

$$\approx \frac{S_0}{\beta} + \frac{1}{2\beta} \sum_{\mathbf{q}, \omega_n} \ln\left[\det M(\mathbf{q}, i\omega_n)\right],$$

$$= \Omega_0 + \Omega_f \tag{9.40}$$

with Ω_0 denoting the thermodynamic potential from the mean-field calculation.

From here, it is straightforward to derive the gap and number equations. For the number equation,

$$n = -\frac{\partial \Omega}{\partial \mu}$$

$$= -\frac{\partial \Omega_0}{\partial \mu} - \frac{\partial \Omega_f}{\partial \mu}. \tag{9.41}$$

Note the second term in the last line above corresponds to the contribution due to the Gaussian fluctuations, while the first term corresponds to contributions on the mean-field level. For the gap equation, a common practice is to let it take the mean-field form

$$\frac{\partial \Omega_0}{\partial \Delta_0^*} = 0. \tag{9.42}$$

The physical argument behind the choice is that by preserving the mean-field form of the gap equation, one gets a description in which the low-energy excitations are consistent with the Goldstone theorem in the broken symmetry phase below T_c. Indeed, it has been shown that if one uses a full gap equation

$$\frac{\partial \Omega}{\partial \Delta_0^*} = 0, \qquad (9.43)$$

together with the number equation, the result is not improved [3, 5].

9.3. Extension of the NSR scheme based on the T-matrix formalism

Another widely applied school of approaches for the beyond-mean-field calculations is the T-matrix formalism. In this section, we introduce the basic concepts of these schemes based on the two-body scattering T-matrix. As our discussion on the NSR is based on two-body scattering, the T-matrix formalism can also be viewed as natural extensions of the original NSR scheme.

We start from a normal attractive Fermi gas that is above the critical temperature, where it is easy to construct the Dyson's series Eq. (9.3) from the two-body scattering T-matrix. In the original NSR scheme, the Dyson's equation for a single particle Green's function is truncated at the lowest order. Diagrammatically, terms corresponding to those in Fig. 9.3 are not included in the summation. Hence, a straightforward extension of the NSR scheme is to include these higher-order terms by adopting the full Dyson's equation

$$G^{-1} = G_0^{-1} - \Sigma, \qquad (9.44)$$

where the self-energy Σ is determined from the T-matrix and the single particle Green's function for free fermions as in Eq. (9.2).

In the NSR scheme, the T-matrix above the critical temperature is calculated using

$$T(Q) = U - U\chi(Q)T(Q), \qquad (9.45)$$

with the pair susceptibility χ composed of free fermion propagators

$$\chi(Q) = \frac{1}{\beta} \sum_K G_0(K)G_0(Q - K). \qquad (9.46)$$

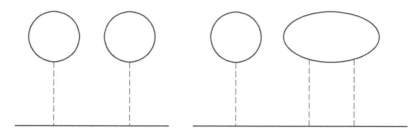

Figure 9.3.   Illustration of the Feynman diagrams not included in the NSR scheme.

One may also consider the approach where the free fermion Green's functions are replaced by the full Green's function

$$\chi(Q) = \frac{1}{\beta} \sum_K G(K)G(Q - K). \tag{9.47}$$

Correspondingly, the self-energy should be

$$\Sigma(K) = \frac{1}{\beta} \sum_Q T(Q)G(Q - K). \tag{9.48}$$

Together with the Dyson's equation, these equations self-consistently determine the thermodynamic potential.

Along this way of thinking, one may also take an asymmetric form of the propagator

$$\chi(Q) = \frac{1}{\beta} \sum_K G(K)G_0(Q - K), \tag{9.49}$$

while the self-energy is still determined using free fermion propagators as in Eq. (9.2).

In all these different variations of the NSR scheme, the onset of superfluidity is characterized by the appearance of a pole in the T-matrix. The critical temperature T_c can be calculated via the corresponding Thouless condition $1 + U\chi(0) = 0$. These schemes have all been discussed extensively in the literature. While the critical temperatures calculated using different schemes typically differ slightly from one another, they are all considerably lower than that of the mean-field result. This is consistent with the expectation that fluctuations tend to destroy the long-range order in the superfluid phase.

At the critical temperature T_c, the attractively interacting Fermi gas undergoes a phase transition into the broken symmetry phase. This is characterized by a pole in the T-matrix, which we have already used to calculate T_c. To extend the NSR scheme below T_c, we must modify the T-matrix and the self-energy such that the physics of pairing superfluidity is accounted for.

Here, as an example, we review a simple and formally elegant scheme proposed by Chen et $al.$, where an asymmetric form of the susceptibility was adopted [4]. To appreciate the motivation of the formalism, let us first recast the standard BCS theory in an asymmetric propagator form.

We start from the single-particle Green's function in the Nambu formalism, where

$$
\begin{aligned}
(\mathcal{G}_0)_{11}(K) &= \frac{i\omega_n + \xi_{\mathbf{k}}}{-\omega_n^2 - E_{\mathbf{k}}} \\
&= \left(i\omega_n - \xi_{\mathbf{k}} - \frac{|\Delta|^2}{i\omega_n + \xi_{\mathbf{k}}} \right)^{-1} \\
&= \left[G_0^{-1} + |\Delta|^2 G_0(-K) \right]^{-1} \\
&= \left[G_0^{-1} - \Sigma_{\mathrm{sc}} \right]^{-1} \\
&= G(K).
\end{aligned}
\tag{9.50}
$$

Note that during the derivation, we have identified the self-energy $\Sigma_{\mathrm{sc}} = -|\Delta|^2 G_0(-K)$. Physically, this corresponds to the virtual process in which two fermions form a pairing state in the condensate, which then breaks up into separate fermions. To be consistent with the T-matrix formalism, this self-energy can be written as

$$
\begin{aligned}
\Sigma_{\mathrm{sc}}(K) &= \frac{1}{\beta} \sum_Q T(Q) G_0(Q - K) \\
&= -\Delta^2 G_0(-K),
\end{aligned}
\tag{9.51}
$$

which suggests that in the broken symmetry phase, the propagator describing the condensed pairs can be written as

$$
T(Q) = -\beta \Delta^2 \delta(Q).
\tag{9.52}
$$

The singularity in the T-matrix here is a consequence of the condensation of the pairs into the state with $Q = 0$.

Substituting the self-energy into the Dyson's equation, we recover the finite temperature mean-field number equation

$$n = \frac{2}{\beta} \sum_K G(K)$$

$$= \frac{2}{\beta} \sum_K \left[G_0^{-1}(K) + |\Delta|^2 G_0(-K) \right]^{-1}$$

$$= \sum_{\mathbf{k}} \left[1 - \frac{\xi_{\mathbf{k}}}{E_{\mathbf{k}}} + 2 \frac{\xi_{\mathbf{k}}}{E_{\mathbf{k}}} f(E_{\mathbf{k}}) \right], \qquad (9.53)$$

where we have performed the Matsubara summation in the last line. The mean-field gap equation on the other hand can be cast into the following form

$$1 + U\chi(0) = 0, \qquad (9.54)$$

with an asymmetric pair propagator $\chi(Q) = \frac{1}{\beta} \sum_K G(K)G_0(Q-k)$. More explicitly,

$$-\frac{1}{U} = \chi(0)$$

$$= \frac{1}{\beta} \sum_K \left[G_0^{-1}(K) + |\Delta|^2 G_0(-K) \right]^{-1} G_0(-K)$$

$$= \frac{1}{\beta} \sum_{\mathbf{k},\omega_n} \left[i\omega_n - \xi_{\mathbf{k}} + \frac{|\Delta|^2}{-i\omega_n - \xi_{\mathbf{k}}} \right]^{-1} \frac{1}{-i\omega_n - \xi_{\mathbf{k}}}$$

$$= \sum_{\mathbf{k}} \frac{1}{2E_{\mathbf{k}}} \tanh \frac{\beta E_{\mathbf{k}}}{2}. \qquad (9.55)$$

Hence the finite-temperature gap equation for $T < T_c$ can be interpreted as the critical condition for the non-condensed pairs, which corresponds to a pole at zero energy/momentum in the single fermion T-matrix

$$T(Q) = \frac{U}{1 + U\chi(Q)}, \qquad (9.56)$$

where $\chi(Q)$ takes the asymmetric form.

To incorporate thermal fluctuations into the formalism above, we need to include the self-energy induced by non-condensed pairs. In contrast to contribution from the condensed pairs, the T-matrix corresponding to non-condensed pairs has non-zero energy/momentum. We follow the notation in Ref. [4] and write the total self-energy as

$$\Sigma(K) = \Sigma_{sc} + \Sigma_{pg}$$

$$= -\Delta^2 G_0(-K) + \frac{1}{\beta}\sum_{Q\neq 0} T_{pg}(Q)G_0(Q-K), \qquad (9.57)$$

where the subscript in T_{pg} stands for a pseudogap. As we will see in the following, the fluctuation-induced self-energy Σ_{pg} corresponds to the existence of an excitation gap that persists even above the critical temperature T_c.

One critical assumption of the asymmetric scheme discussed here is that the propagator for the non-condensed pairs $T_{pg}(Q)$ should be sharply peaked at $Q = 0$. In this case, the total self-energy can also be written in a BCS form (Eq. (9.51)), thus providing a dramatic simplification

$$\Sigma(K) \approx -(\Delta^2 + \Delta^2_{pg})G_0(-K)$$

$$= -\tilde{\Delta}^2 G_0(-K), \qquad (9.58)$$

with $\tilde{\Delta}^2 = \Delta^2 + \Delta^2_{pg}$ and the pseudogap

$$\Delta^2_{pg} = \frac{1}{\beta}\sum_{Q\neq 0} T_{pg}(Q). \qquad (9.59)$$

Consistent with the physical picture of a pseudogap, the excitation gap $\tilde{\Delta}$ has contributions from both the condensed and the non-condensed pairs. As the temperature goes up from zero, the contribution from the non-condensed pairs Δ_{pg} becomes larger as that from the condensed pairs Δ decreases. At the critical temperature T_c, Δ vanishes while Δ_{pg} is still finite. This pseudogap only becomes zero at a much higher temperature, at which point the fermions no longer form pairs. From this understanding, the total gap $\tilde{\Delta}$ appearing in the gap equation is no longer the order parameter at finite temperatures. To characterize the critical temperature for the condensation of fermion pairs, one needs to evaluate the pseudogap Δ_{pg}.

The pseudogap above can be determined by writing

$$\chi(Q) = \frac{1}{\beta}\sum_{K} G(-K)G_0(Q+K)$$

$$\approx \frac{1}{\beta}\left[G_0^{-1}(-K) + \Delta^2 G_0(K)\right]^{-1} G_0(Q+K). \qquad (9.60)$$

It is then straightforward to perform the Matsubara summation and get a closed-form expression for $\chi(Q)$. The structure of T_{pg} can be made clearer by expanding the pair susceptibility $\chi(Q)$ to quadratic order in Q

$$T_{\text{pg}} = \frac{U}{1 + U\chi(Q)}$$

$$\approx \frac{1}{i\Omega_n - \frac{\hbar q^2}{2M^*}}, \tag{9.61}$$

with M^* determined by the expansion coefficient. This implies that with the assumption of the sharp structure in $T_{\text{pg}}(Q)$, the pseudogap is induced by the propagation of bosonic quasi-particle excitations with an effective mass M^*. Note that unlike the previous schemes, the low-energy excitations caused by the pairing fluctuation is quadratic, corresponding to quasi-particle excitations. This is in contradiction to the existence of Goldstone modes at low energies for the broken symmetry phase, and is an artifact of the assumption on the form of the self-energy.

The fluctuations evaluated in this way would modify the BCS number equation by adding a term accounting for the population of the non-condensed fermion pairs

$$n_b = 2 \sum_{\mathbf{q}} \frac{1}{e^{\hbar^2 q^2/2M^*} - 1}. \tag{9.62}$$

Thus, all physical parameters can be evaluated by self-consistently solving the gap equation, the modified number equation and Eq. (9.59). In the zero-temperature limit, the equations of the asymmetric propagator scheme reduce to the mean-field equations. This implies that all the bosonic quasi-particles condense at zero temperature, which is consistent with the physical picture discussed at the beginning.

The advantage of the asymmetric propagator formalism reviewed here is the natural and transparent inclusion of the pseudogap Δ_{pg}. The scheme is also numerically less demanding than its counterparts with symmetric propagators. We note that the formalism can be derived more formally from the equations of motion for the Green's functions.

References

[1] Nozières and S. Schmitt-Rink. *J. Low Temp. Phys.* **59**, 195 (1985).
[2] V. M. Galitskii. *Zh. Eksp. Teor. Fis.* **34**, 151 (1958). [*Sov. Phys. JETP* **7**, 104 (1958)].

[3] R. B. Diener, R. Sensarma and M. Randeria. *Phys. Rev. A* **77**, 023626 (2008).

[4] Q. Chen, J. Stajic, S. Tan and K. Levin. *Phys. Rep.* **412**, 1 (2005).

[5] P. Pieri, L. Pisani and G. C. Strinati. *Phys. Rev. B* **72**, 012506 (2005).

[6] C. A. R. Sá de Melo, M. Randeria and J. R. Engelbrecht. *Phys. Rev. Lett.* **71**, 3202 (1993).

[7] J. W. Negele and H. Orland. *Quantum Many-particle Systems*. Westview (1998).

[8] G. D. Mahan. *Many-particle Physics*. Kluwer Academic/Plenum Publishers, New York (2000).

10
Polarized Fermi Gas

In the previous chapters, we have discussed resonant s-wave superfluidity in Fermi gases with an equal population in the two spin states. In these systems, the Fermi surfaces of the different spin species overlap. In the weak-coupling limit, the Fermi surface becomes unstable due to the Cooper instability in the presence of a weak attractive interaction, and fermions with different spin species and on opposite sides of the Fermi surface tend to form Cooper pairs. On the other hand, for a Fermi gas where the Fermi surfaces of the two spin species do not overlap, e.g., in a spin-polarized Fermi gas or in a Fermi gas with atoms of different masses, the above scenario may fail to apply. Due to the mismatch of the Fermi surface, Cooper pairs are formed at an energy cost. In general, this implies that the ground state may be different from the standard BCS pairing state and a quantum phase transition is possible as the difference in the Fermi surfaces increases.

Historically, the pairing superfluidity of spin-polarized fermions has been studied in different physical contexts, mostly in the weakly interacting regime. In addition to the existence of a quantum phase transition between the superfluid state and the normal state, various exotic phases have been proposed to exist due to the competition between the Cooper pairing and the chemical potential imbalance. However, these phases and phase transitions have not been observed before in conventional materials, such as superfluid ^{3}He and superconducting metals. In these systems, the chemical potential difference is usually induced by an external magnetic field. The magnitude of the magnetic field, however, cannot be made large as it would create vortices in the bulk and eventually destroy the superfluid ground state. Another exemplary physical system is the neutron star, where like in the nuclear matter, quarks of different flavors and masses condense into a superfluid of Cooper pairs. The characterization of the novel phases and the phase transitions in a polarized Fermi gas therefore has far-reaching implications in understanding the properties of these systems in completely different physical contexts.

Toward the end of 2005, the MIT and Rice research groups almost simultaneously reported the observation of quantum phase transitions in ultracold Fermi gases near wide Feshbach resonances that have imbalanced spin populations [1, 2]. In ultracold Fermi gases, the atoms live in a much cleaner environment and the Feshbach resonance technique provides a convenient knob for tuning the s-wave interaction between the atoms. Additionally, the chemical potential difference necessary for the observation of interesting phases and phase transitions can be easily implemented by adjusting the population of atoms in each hyperfine state. Thus, the experimentalists were able to study the effects of the competition between pairing and chemical potential difference throughout the crossover region in a controlled way. These experiments created a good deal of excitement and stimulated many studies on polarized Fermi gases, as they provide an unparalleled opportunity for generating and characterizing exotic quantum phases with ultracold atoms.

In this chapter, we discuss various interesting aspects related to a polarized Fermi gas. We will focus on the mean-field description here. Simple as it is, the mean-field theory is sufficient to capture most of the interesting physics herein. However, it is also important to point out that while the experimental observations can be explained qualitatively with the mean-field calculations, the mean-field results are often incorrect quantitatively, e.g., it always overestimates the quantum critical point. We note that it is straightforward to extend the beyond-mean-field formalisms that we discussed in the previous chapter to the polarized case. We also investigate the possible existence of exotic quantum phases and examine in more detail its normal state in the presence of an attractive interaction with large chemical potential difference.

10.1. Mean-field results

For experimental relevance, we restrict our discussion to a wide Feshbach resonance and rewrite the Hamiltonian in Eq. (10.1) to get

$$H - \mu_\uparrow N_\uparrow - \mu_\downarrow N_\downarrow = \sum_{\mathbf{k},\sigma} (\epsilon_\mathbf{k} - \mu_\sigma) a^\dagger_{\mathbf{k},\sigma} a_{\mathbf{k},\sigma}$$

$$+ (1/\mathcal{V}) \sum_{\mathbf{k},\mathbf{k}'} U_{\mathbf{k},\mathbf{k}'} a^\dagger_{\mathbf{k},\uparrow} a^\dagger_{-\mathbf{k},\downarrow} a_{-\mathbf{k}',\downarrow} a_{\mathbf{k}',\uparrow}. \qquad (10.1)$$

Here we again assume a contact interaction potential between different spin species, such that $U_{\mathbf{k},\mathbf{k}'} = U$, where the bare interaction rate U should be renormalized as before.

The order parameter can be defined in the same way as in the unpolarized case: $\Delta = -U/\mathcal{V} \sum_\mathbf{k} \langle a_{-\mathbf{k},\downarrow} a_{\mathbf{k},\uparrow} \rangle$. One may then follow the same procedure in diagonalizing the effective BCS Hamiltonian to get

$$H_{\text{eff}} - \sum_\sigma \mu_\sigma N_\sigma = \sum_{\mathbf{k},\lambda=\pm} E_{\mathbf{k},\lambda} \alpha^\dagger_{\mathbf{k},\lambda} \alpha_{\mathbf{k},\lambda} + \sum_\mathbf{k} [\epsilon_\mathbf{k} - \mu]$$

$$- \frac{1}{2} \sum_{\mathbf{k},\lambda} E_{\mathbf{k},\lambda} - \mathcal{V} \frac{\Delta^2}{U}. \qquad (10.2)$$

The chemical potentials $\mu = (\mu_\uparrow + \mu_\downarrow)/2$ and $h = (\mu_\uparrow - \mu_\downarrow)/2$ can be tuned by changing the population in the two different hyperfine spin states. Compared with the unpolarized gas, the most apparent difference is the splitting of the excitation spectrum in the case of a polarized Fermi gas. The two different branches of the dispersion spectrum are

$$E_{\mathbf{k},\pm} - \sqrt{|\Delta|^2 + (\epsilon_\mathbf{k} - \mu)^2} \pm h. \qquad (10.3)$$

In contrast to that of an unpolarized Fermi gas, $E_{\mathbf{k},\pm}$ are not positive semidefinite here, but rather dependent on actual physical parameters. When the energy of quasi-particle excitations $E_{\mathbf{k},\lambda}$ become negative, the vacuum of quasi-particles are not the ground state anymore and one needs to perform a particle-hole transformation $\{\alpha^\dagger_{\mathbf{k},\lambda} \to \alpha_{\mathbf{k},\lambda}, \alpha_{\mathbf{k},\lambda} \to \alpha^\dagger_{\mathbf{k},\lambda}\}$ in the corresponding region in momentum space to find the new ground state. To make this point clearer, we require the dispersion spectra to be positive and rewrite the effective Hamiltonian to get

$$H_{\text{eff}} - \sum_{\sigma} \mu_{\sigma} N_{\sigma} = \sum_{\mathbf{k},\lambda=\pm} |E_{\mathbf{k},\lambda}| \alpha_{\mathbf{k},\lambda}^{\dagger} \alpha_{\mathbf{k},\lambda} + \sum_{\mathbf{k}} [\epsilon_{\mathbf{k}} - \mu]$$

$$+ \sum_{\mathbf{k},\lambda} \left[\theta(-E_{\mathbf{k},\lambda}) E_{\mathbf{k},\lambda} - \frac{E_{\mathbf{k},\lambda}}{2} \right] - \mathcal{V} \frac{\Delta^2}{U}. \quad (10.4)$$

As the Hamiltonian is symmetric with respect to a spin flip, the physics would be the same regardless of which spin species is the majority component. Without loss of generality, we will take $h > 0$ in the following discussions.

For $h > 0$, the excitation for the branch with $E_{\mathbf{k},+}$ is always positive for non-vanishing order parameters $\Delta \neq 0$. The quasi-particle excitation of this branch is therefore gapped, just as in the typical BCS theory for an unpolarized gas. Therefore, one must overcome a finite energy gap to excite a quasi-particle of this branch. However, for the other branch, the excitation energy $|E_{\mathbf{k},-}|$ may become zero even in the presence of a non-vanishing order parameter. Consider the condition for $E_{\mathbf{k},-} = 0$,

$$(\epsilon_{\mathbf{k}} - \mu)^2 + |\Delta|^2 = h^2. \quad (10.5)$$

When Eq. (10.5) has real solutions in momentum space, the quasi-particle excitations of the $E_{\mathbf{k},-}$ branch with the corresponding momenta cost zero energy. In other words, the excitation gap becomes zero in the presence of a non-vanishing pairing gap. This remarkable behavior signals the existence of a novel superfluid phase with gapless excitations. The important question then is the stability of this non-trivial superfluid phase. For this purpose, let us examine the thermodynamic potential.

At a finite temperature T, the thermodynamic potential can be written as

$$\Omega = -T \ln[\text{tr}(e^{-(H - \mu_{\uparrow} N_{\uparrow} - \mu_{\downarrow} N_{\downarrow})/T})],$$

$$= -\Delta^2 \mathcal{V}/U_T - T \sum_{\mathbf{k}} \{ \ln[1 + \exp(-|E_{\mathbf{k}\downarrow}|/T)]$$

$$+ \ln[1 + \exp(-E_{\uparrow}/T)] - [\epsilon_{\mathbf{k}} - \mu_{\uparrow} - \theta(E_{\downarrow}) E_{\downarrow}]/T \}. \quad (10.6)$$

Intuitively, one may follow the standard treatment and apply the extrema condition $\partial\Omega/\partial\Delta = 0$ and the number constraint $N_{\sigma} = -\partial\Omega/\partial\mu_{\sigma}$ to derive the gap and the number equations. In the case of a polarized Fermi gas, however, there is one further complication. Due to the competition

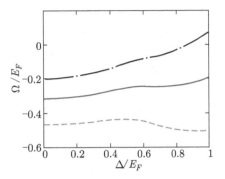

Figure 10.1. Typical structure of the thermodynamic potential as a function of the pairing gap. The chemical potential μ increases from the top to the bottom, while h is fixed. The unit of energy E_F is defined as $E_F = \hbar^2(3\pi^2 n)^{2/3}/2m$, where n is the total density.

between the pairing superfluidity and the chemical potential difference, the solutions of the gap equation may correspond to meta-stable or unstable states, besides the ground state. We can see it more clearly in Fig. 10.1, where the thermodynamic potential is plotted as a function of Δ for various chemical potentials. Unlike the unpolarized case, the thermodynamic potential may develop a double well structure within a certain range of parameters. As the solution of the gap equation only corresponds to local extrema, one should look instead for the global minimum of the thermodynamic potential.

Furthermore, for a uniform system, it is not guaranteed that one can find solutions of the gap equation and number equations simultaneously at the global minimum of the thermodynamic potential. Under these circumstances, the ground state of the system should be a phase-separated state, in which two different phases co-exist in a uniform gas. From detailed numerical calculations of the thermodynamic potential, it can be concluded that the phase-separated state in a polarized Fermi gas typically consists of two different phases: a normal (N) and a superfluid (SF) phase. In principle, phase-separation consisting of more exotic phases may occur, however, their stability regions are usually limited in a three-dimensional gas. Hence, we will only discuss the phase-separated state of the N and SF phases.

To characterize the phase-separated state, one needs to introduce an additional mixing coefficient $0 \leq x \leq 1$ to the thermodynamic potential [3]

$$\Omega = x\Omega(\Delta) + (1 - x)\Omega(0), \qquad (10.7)$$

where $\Omega(\Delta)$ stands for the contribution from the SF phase with non-vanishing order parameter Δ and $\Omega(0)$ the contribution from the N phase. When $x = 0$, Eq. (10.7) gives the thermodynamic potential of a pure SF phase and when $x = 1$, it describes a pure N phase. For a phase-separated state, the number equation should be modified as

$$n_\sigma = -x \frac{\partial \Omega(\Delta)}{\partial \mu_\sigma} - (1 - x) \frac{\partial \Omega(0)}{\partial \mu_\sigma}, \qquad (10.8)$$

where Δ corresponds to a local minimum of the thermodynamic potential. Finally, we have the degeneracy equation $\partial \Omega / \partial x = 0$ for the phase-separated state

$$\Omega(\Delta) = \Omega(0). \qquad (10.9)$$

With these considerations, we can now solve the ground state for a uniform polarized Fermi gas and map out its phase diagram.

Before discussing the system's phase diagram, let us focus on how the population difference is accounted for. From the number equations, the population difference between the two spin species can be derived

$$\delta n = n_\uparrow - n_\downarrow = -\frac{\partial \Omega}{h}. \qquad (10.10)$$

In the BCS SF state, the excitation spectrum is gapped, i.e., $E_{\mathbf{k},\lambda} > 0$. From Eq. (10.6), the population difference in a BCS SF phase can only be carried by thermal excitations at a finite temperature

$$\delta n = f(E_{\mathbf{k},-}) - f(E_{\mathbf{k},+}). \qquad (10.11)$$

In the N state, $\Delta = 0$, and under our mean-field description, the system is reduced to two non-interacting Fermi seas with different Fermi surfaces

$$n_\sigma = \sum_{\mathbf{k}} f(\epsilon_{\mathbf{k}} - \mu_\sigma). \qquad (10.12)$$

Finally, if the system is in a gapless SF phase as we discussed before, i.e., $E_{\mathbf{k},-} < 0$, the number difference becomes

$$\delta n = \sum_{\mathbf{k}} [1 - f(E_{\mathbf{k},+}) - f(E_{\mathbf{k},-})]. \qquad (10.13)$$

From the analysis above, we see that in the zero-temperature limit, the population difference is carried either by the N state, in which each spin species fills the corresponding Fermi sea, or by the region in momentum space satisfying $E_{\mathbf{k},\lambda} < 0$ for the non-trivial SF phase. We plot the momentum distribution for typical gapless SF states in Fig. 10.2. At zero temperature, these novel quantum phases can be regarded as phase separation in momentum space, with one (BP1, where BP stands for breached pair) or two (BP2) Fermi surfaces [5, 8]. The polarization is carried by the "normal phase" in the momentum space. Note that the order parameter Δ is always finite. A detailed calculation shows that in a three-dimensional polarized Fermi gas, the BP2 phases is not stable while the BP1 phase can be stabilized on the BEC side of the crossover.

Figure 10.3 shows the typical phase diagram for a uniform polarized Fermi gas. Near the unitary region, the ground state of a polarized Fermi gas

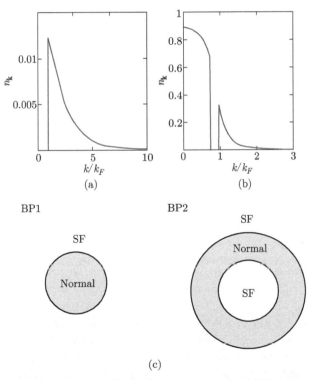

Figure 10.2.   Characterization of the breached pairing states. (a) The momentum distribution of the minority spin component in a BP1 phase. (b) The momentum distribution of the minority spin component in a BP2 phase. (c) Schematics for the momentum space phase separation for the breached pairing states.

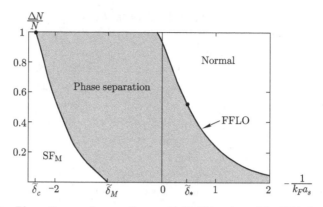

Figure 10.3.   Phase diagram for a uniform polarized Fermi gas. The BP1 phase is labeled SF_M in the phase diagram and the stability region of the FFLO phase is also indicated. From Ref. [3].

is a phase-separated state. The stability region of the phase-separated state becomes exponentially small in the BCS limit, where beyond a vanishingly small polarization, the superfluidity of the system would be lost and the Fermi gas becomes normal. On the BEC side, however, there is a large stability region for the BP1 phase.

From the analysis above, we see that the BCS SF phase only accommodates population imbalance at finite temperatures. At zero temperature and finite population imbalance, the system cannot be in a pure BCS gapped SF state. Instead, the ground state is driven into a phase-separated state or a BP1 phase that has built-in polarization.

10.2.   Fulde–Ferrell–Larkin–Ovchinnikov (FFLO) phase

Near the unitary region and on the BCS side, the phase-separated state undergoes a phase transition as the polarization is increased. Near the border of the phase transition, the order parameter of the SF component vanishes and the mean-field theory becomes unreliable, in the sense that the ground state may no longer be in the mean-field BCS form. Indeed, on the BCS side of the resonance, people have identified another type of novel quantum phase, the so-called FFLO phase, which also helps to carry the population imbalance.

The FFLO phase was first studied in the context of condensed matter physics [6, 7]. It was suggested that the pairing state in a BCS superconductor under a constant magnetic field might acquire a non-zero center-of-mass momentum $\mathbf{Q}$. The corresponding order parameter is then modified by a spatially dependent phase factor

$$\Delta_{\mathbf{Q}} = -U/\mathcal{V} \sum_{\mathbf{k}} \left\langle a_{\mathbf{Q}/2-\mathbf{k},\downarrow} a_{\mathbf{Q}/2+\mathbf{k},\uparrow} \right\rangle = \Delta e^{-i\mathbf{Q}\mathbf{r}}. \qquad (10.14)$$

It was later suggested that a multi-$\mathbf{Q}$ FFLO phase is in fact more stable than the simple single-$\mathbf{Q}$ phase, although the energy difference between them is rather small. This implies a spatially periodically modulated order parameter. Despite the efforts over the years, unequivocal evidence for the FFLO phase has not yet been established in condensed matter systems.

The successful preparation of a polarized Fermi gas in the strongly interacting regime rekindled the hope of the experimental observation of the FFLO state, this time in an ultracold Fermi gas. Various theoretical proposals have been put forth on the preparation and observation of the phase. For a uniform polarized Fermi gas in three spatial dimensions, it has been reported by several research groups that the FFLO phase is only stable in a narrow parameter window on the BCS side of the resonance, bordering the phase-separated state and the N state. In lower dimensions, on the other hand, the FFLO phase can be stabilized in a larger parameter range. Here, we will outline the characterization of a FFLO state in the simplest mean-field framework in three dimensions and briefly discuss the resulting phase diagram.

The FFLO phase that we discuss here can be understood as a condensate of pairing states with a non-zero center-of-mass momentum. As we have seen in the previous section, the competition between pairing and chemical potential difference may lead to quantum phase transitions between SF and N phases. Another way to compromise this competition is to have a pairing state with the Fermi surfaces shifted, i.e., to acquire a non-zero center-of-mass momentum. This is equivalent to taking a new form of the BCS-type many-body wave function

$$|\Psi\rangle = \prod_{\mathbf{k}} \left(u_{\mathbf{k}} + v_{\mathbf{k}} a^{\dagger}_{\mathbf{k}+\frac{\mathbf{Q}}{2},\uparrow} a^{\dagger}_{\frac{\mathbf{Q}}{2}-\mathbf{k},\downarrow} \right) |\text{vac}\rangle. \qquad (10.15)$$

Here we only consider the so-called Fulde–Ferrell phase, in which the pairing states acquire a single center-of-mass momentum $\mathbf{Q}$. The order parameter also needs to be modified as in Eq. (10.14).

Following the mean-field formalism and diagonalizing the effective
Hamiltonian, we may derive two different branches of quasi-particle excitations with the dispersion relations

$$E_{\mathbf{k},\pm} = \sqrt{(\epsilon_{\mathbf{k}} + \epsilon_{\frac{\mathbf{Q}}{2}} - \mu)^2 + |\Delta_{\mathbf{Q}}|^2} \pm \left(\frac{\mathbf{k} \cdot \mathbf{Q}}{2m} - h \right), \qquad (10.16)$$

where $\epsilon_{\frac{\mathbf{Q}}{2}} = \hbar^2 Q^2/8m$. Clearly, the quasi-particle excitation can become
gapless at specific points in the momentum space.

Typically, there may exist several local minima on the thermodynamic
potential landscape on the plane of Δ and Q [9], which respectively correspond to the N state, the BCS SF state and the FFLO state. To investigate
the stability of the FFLO phase with respect to other possible phases and
phase-separated states, one needs to calculate the energy of different scenarios and find a state that minimizes the thermodynamic potential. This
complicated task has been carried out by several research groups and the
resulting phase diagram (Fig. 10.3) shows that the FFLO phase that we
consider is only stable in a narrow parameter region on the BCS side and
on the border of the N and phase-separated states. Note that this is the
zero-temperature result at the mean-field level. At finite temperatures and
with beyond-mean-field calculations, as thermal and quantum fluctuations
can accommodate part of the population imbalance, we may expect the
stability region of the FFLO phase to shrink futher, thus making it even
more difficult to be observed experimentally. On the other hand, as the
stability region of the FFLO state becomes larger at lower dimensions,
signatures of FFLO states are typically easier to detect in low dimensional
Fermi gases. Indeed, in a recent seminal experiment, the Rice research group
has identified strong evidence for an FFLO state in a quasi-one-dimensional
Fermi gas in an optical lattice [11].

10.3. Polarized Fermi gas in a trap

So far we have characterized the phase diagram in a uniform system. Experimentally, the Fermi gases are always confined in a trapping potential, which
necessarily makes the system anisotropic. It is therefore important to understand the phase configurations in a trapping potential. One may of course
apply the BdG formalism in this new setting and people have done exactly
that [12]. For the discussion here, we will adhere to the LDA approach,
for simplicity and for physical transparency. However, we emphasize that

the results from the LDA calculations are only valid when the trapping potential varies slowly in space and that the number density in the center of the trap is sufficiently large.

Under the LDA, each spatial location can be considered to be infinitely large and the local chemical potential is connected to that at the center of the trap by the simple relation

$$\mu_\sigma(\mathbf{r}) = \mu_\sigma - V(\mathbf{r}). \tag{10.17}$$

Here $V(\mathbf{r})$ represents the trapping potential. It is easy to see that the chemical potential difference h is constant throughout the trap and $\mu(\mathbf{r}) = \mu_0 - V(\mathbf{r})$. As in the case of an unpolarized gas, the chemical potentials h and μ_0 can be solved by imposing the particle number constraints

$$N = \int d^3\mathbf{r}[n_\uparrow(\mathbf{r}) + n_\downarrow(\mathbf{r})], \tag{10.18}$$

$$P = \int d^3\mathbf{r} \frac{n_\uparrow(\mathbf{r}) - n_\downarrow(\mathbf{r})}{N}, \tag{10.19}$$

where $N = N_\uparrow + N_\downarrow$ is the total particle number and $P = (N_\uparrow - N_\downarrow)/N$ is the total polarization in the trap.

As we have discussed before, the existence of a uniform phase-separated state is necessitated by the fact that the number equations do not have real solutions as no stable pure phase can accommodate the population imbalance alone. Now that the number constraint is on the integrated trap densities, number equations no longer need to be satisfied locally. Therefore, pure phases can be stabilized locally and the Fermi gas naturally phase separates in real space. Figure 10.4 shows a typical phase structure in a trapping potential. For an isotropic trapping potential, different phases form a shell structure, typically with a SF at the center and polarized normal or mixed normal phases appearing as rings toward the edge of the trap. The sharp edges in some of the subplots indicate first-order phase transitions, where the global minimum of the local thermodynamic potential jumps from one local minimum to another. In reality, densities should vary smoothly in space. Therefore, these sharp edges are an artefact of the LDA, which are smoothed out in a BdG calculation. Finally, the FFLO state also shows up in a narrow parameter region as a thin shell between the SF core and normal edge on the BCS side of resonance.

This simple picture of a phase separation in a trap has been observed experimentally. Similar to the unpolarized case, the mean-field description

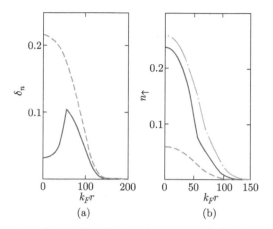

(a)   (b)

Figure 10.4.   The atom density distributions in a trap, calculated with parameters corresponding to the MIT experiment [1] with $T \sim 0.1 E_F$ and $B - B_0 \sim 56.3$G. The solid, the dashed and the dash-dot curves correspond to the imbalance ratio $P \sim 0.46$, $P \sim 0.86$ and $P = 0$, respectively, with the total atom number $N \sim 2.7 \times 10^7$ in all the cases. From Ref. [10].

paints a qualitatively correct picture of phase separation, but fails to provide a quantitatively satisfactory fit with the experimental measurements. In particular, the SF-N phase transition at the center of the trap is reported to occur near $P \approx 0.7$ at unitarity, while the critical polarization is above 0.99 from a mean-field calculation. In this sense, the critical polarization from the mean-field calculation can be viewed as an upper bound for the experimental measurements, as fluctuations will necessarily lower the critical polarization from the mean-field theory.

References

[1] M. W. Zwierlein, A. Schirotzek, C. H. Schunck and W. Ketterle. *Science* **311**, 492 (2006).
[2] G. B. Partridge, W. Li, Y. A. Liao, R. G. Hulet, M. Haque and H. T. C. Stoof. *Phys. Rev. Lett.* **97**, 190407 (2006).
[3] D. E Sheehy and L. Radzihovsky. *Ann. Phys.* **322**, 1790 (2007).
[4] A. M. Clogston. *Phys. Rev. Lett.* **9**, 266 (1962).
[5] G. Sarma, *J. Phys. Chem. Solids* **24**, 1029 (1963).
[6] P. Fulde and R. A. Ferrell. *Phys. Rev.* **135**, A550 (1964).
[7] A. I. Larkin and Yu. N. Ovchinnikov. *Zh. Eksp. Teor. Fiz.* **47**, 1136 (1964) [Sov. Phys. JETP **20**, 762 (1965)].
[8] W. V. Liu and F. Wilczek. *Phys. Rev. Lett.* **90**, 047002 (2003).

 [9] H. Hu and X.-J. Liu. *Phys. Rev. A* **73**, 051603(R) (2006).

[10] W. Yi and L.-M. Duan. *Phys. Rev. A* **73**, 031604(R) (2006).

[11] Y. Liao, A. Sophie, C. Rittner, T. Paprotta, W. Li, G. B. Partridge, R. G. Hulet, S. K. Baur and E. J. Mueller. *Nature* **467**, 567 (2010).

[12] X.-J. Liu, H. Hu and P. D. Drummond. *Phys. Rev. A* **75**, 023614 (2007).

11
Synthetic Gauge Field

The recent experimental realization of a synthetic gauge field in dilute gases of neutral atoms has greatly extended the horizon of quantum simulation in these systems [1–7]. By engineering the atom-laser interaction, one may realize both Abelian or non-Abelian synthetic gauge potentials in ultracold atomic gases. It is then possible to simulate, for example, the behavior of charged particles in an electromagnetic field in a system of neutral atoms [2, 3]. Due to the extraordinary tunability of the synthetic magnetic field in ultracold atoms, it has been suggested that quantum Hall effects can be observed in these systems. Another important example is the implementation of spin-orbit coupling, a non-Abelian gauge field, in cold atomic gases, where the internal degrees of freedom of the atoms are coupled to the center-of-mass motional degrees of freedom [4]. Interesting phenomena like the quantum spin Hall effect, topological insulator and topological superfluidity, where spin-orbit coupling plays a key role, may then be investigated in ultracold atomic gases [8–10]. Furthermore, it is now possible to study the novel effects induced by spin-orbit coupling in an ultracold Bose gas, which do not have counterparts in systems of condensed matter [11, 12]. In this chapter, we present a brief review of the recent experimental and theoretical progresses in this direction, with the emphasis on the novel effects induced by synthetic spin-orbit coupling in an ultracold Fermi gas.

11.1.   Implementing synthetic gauge field

The principle behind most of the proposals for artificial gauge potential is based on the adiabatic theorem and the geometrical phase [13]. In general, by engineering the atom-laser interaction, the atoms feel an adiabatic potential when moving through space. The resulting geometrical phase appearing in the effective Hamiltonian gives rise to an artificial gauge potential. For the most general case, let us consider the following single-particle Hamiltonian

$$H = H_0 + V(\mathbf{r}(t)), \tag{11.1}$$

where $H_0 = \mathbf{p}^2/2m$ is the kinetic energy and $V(\mathbf{r}(t))$ describes the atom-laser coupling, whose spatial dependence is slow-varying in time.

Formally, let us expand the wave function at any given time $|\Psi(\mathbf{r},t)\rangle$ onto the eigen basis $\{|\phi_\alpha(\mathbf{r})\rangle\}$ of $V(\mathbf{r})$

$$|\Psi(\mathbf{r},t)\rangle = \sum_\alpha c_\alpha(\mathbf{r},t)|\phi_\alpha(\mathbf{r})\rangle, \tag{11.2}$$

where c_α is the time-dependent expansion coefficients. Substituting the expansion above into the time-dependent Schrödinger's equation and projecting it into the subspace of the α-th eigenstate, we have

$$i\hbar\frac{\partial}{\partial t}c_\alpha(\mathbf{r},t) = E_\alpha(\mathbf{r})c_\alpha + \sum_\beta \langle\phi_\alpha|H_0 c_\beta(\mathbf{r},t)|\phi_\beta\rangle, \tag{11.3}$$

where $E_\alpha(\mathbf{r})$ satisfies $V(\mathbf{r})|\phi_\alpha(\mathbf{r})\rangle = E_\alpha(\mathbf{r})|\phi_\alpha(\mathbf{r})\rangle$. We consider the special case that the typical energy scale of $V(\mathbf{r})$ is much larger than that of H_0. As a result, the slow motion of the particle adiabatically follows that of the internal dynamics governed by $V(\mathbf{r})$. In this case, we may invoke the adiabatic approximation by retaining only $\beta = \alpha$ in Eq. (11.3) to get

$$i\hbar\frac{\partial}{\partial t}c_\alpha(\mathbf{r},t) = E_\alpha(\mathbf{r})c_\alpha + \langle\phi_\alpha|H_0 c_\alpha|\phi_\alpha\rangle, \tag{11.4}$$

which effectively describes the motion of a particle in the adiabatic potential E_α. To make the geometrical phase stand out, we further examine the term involving the kinetic energy

$$\langle\phi_\alpha|H_0 c_\alpha|\phi_\alpha\rangle = \langle\phi_\alpha|\left(-\frac{\hbar^2}{2m}\right)\nabla^2\left[c_\alpha(\mathbf{r},t)|\phi_\alpha(\mathbf{r})\rangle\right]$$

$$= \frac{1}{2m}\left(-i\hbar\nabla - i\hbar\langle\phi_\alpha|\nabla\phi_\alpha\rangle\right)^2 c_\alpha + \sum_{\beta\neq\alpha}\frac{\hbar^2}{2m}|\langle\phi_\beta|\nabla\phi_\alpha\rangle|^2 c_\alpha,$$

$$= \frac{1}{2m}\left(\vec{p} - \vec{A}\right)^2 c_\alpha + W c_\alpha. \tag{11.5}$$

Here,

$$\vec{A} = i\hbar \langle \phi_\alpha | \nabla \phi_\alpha \rangle \tag{11.6}$$

is the geometrical vector potential and

$$W = \sum_{\beta \neq \alpha} \frac{\hbar^2}{2m} |\langle \phi_\beta | \nabla \phi_\alpha \rangle|^2 c_\alpha \tag{11.7}$$

is the geometrical scalar potential. Hence, apart from an energy shift due to the scalar potential W, the effective Hamiltonian for the particle in the adiabatic potential E_α can be written as

$$H_{\text{eff}} = \frac{1}{2m} \left(\vec{P} - \vec{A} \right)^2 + E_\alpha. \tag{11.8}$$

The physical implication is just what we have stated at the beginning: for a particle moving in an adiabatic potential, its external motion adiabatically follows the internal dynamics at each spatial location due to the geometrical phase. As a result, the internal states of the particle may change as the particle is moving through space. When the change in internal states involves only a phase factor, the gauge potential associated with the geometrical phase is Abelian, which is the case for synthetic electromagnetic fields; when the change in internal states involves a general rotation in the Hilbert space spanned by the internal states, the gauge potential associated with the geometrical phase can be non-Abelian, which is the case for synthetic spin-orbit coupling.

Experimentally, the adiabatic potential is generated by coupling the internal states of atoms with a laser. There have also been various proposals to realize synthetic gauge potentials using dark states in multi-pod level schemes or in lattices via Raman-assisted hopping, so on and so forth [13]. Experimentally, it was Spielman's group at the National Institute of Standards and Technology (NIST) that first realized a uniform vector gauge potential in a BEC of ^{87}Rb atoms [1]. The scheme utilizes Raman lasers to couple hyperfine states in the ground state manifold of ^{87}Rb, such that when the atom jumps from one internal state to another via the Raman process, its center-of-mass momentum also changes. Applying a similar scheme, vortices were later observed in a BEC of ^{87}Rb atoms with a synthetic magnetic field. A synthetic electric field was also implemented in an ultracold gas of neutral atoms [2, 3]. In 2010, via a slightly modified Raman scheme, Spielman's research group was able to generate spin-orbit coupling,

a non-Abelian version of the synthetic gauge field, in a BEC of ^{87}Rb atoms
[4, 14, 15]. This was soon followed by the realization of synthetic spin-orbit
coupling in ultracold Fermi gases by the ShanXi and MIT research groups
in 2012 [5, 6].

In the NIST experiments for synthetic electromagnetic fields, the hyper-
fine states $|F, m_F\rangle = \{|1, \pm 1\rangle, |1, 0\rangle\}$ were coupled via counterpropagating
Raman lasers, such that the single-particle Hamiltonian under the rotating-
wave approximation can be written in the hyperfine basis $\{|-1\rangle, |0\rangle, |1\rangle\}$

$$
H = \begin{pmatrix} \frac{\hbar^2 k^2}{2m} - \delta & \frac{\Omega}{2} e^{i2k_0 x} & 0 \\ \frac{\Omega}{2} e^{-i2k_0 x} & \frac{\hbar^2 k^2}{2m} - \epsilon & \frac{\Omega}{2} e^{2ik_0 x} \\ 0 & \frac{\Omega}{2} e^{-2ik_0 x} & \frac{\hbar^2 k^2}{2m} + \delta \end{pmatrix}, \tag{11.9}
$$

where Ω is the effective Rabi frequency of the Raman process, δ is the two-
photon detuning and k_0 is the wave vector of the Raman lasers. Following
the general protocol for evaluating the geometrical phase, the vector gauge
potential in this case lies along the x-axis. As the adiabatic potential can
be rendered either spatial dependent or temporal dependent, a synthetic
magnetic field or electric field can be subsequently realized.

In the scheme above, only the lowest adiabatic potential is considered. As the particle moves along the adiabatic potential, the geometrical
phase accumulates along the path $\gamma = \int \vec{A} d\mathbf{r}$, where $\vec{A}$ is the vector gauge
potential. In contrast, if the Hamiltonian of the system involves multiple
adiabatic potentials, e.g., interaction between particles with different internal states, the effective gauge potential can be non-Abelian. An outstanding
example is the synthetic spin-orbit coupling that has been realized by the
NIST group in 2010, in a spinor BEC of two internal states. The scheme can
actually be viewed as an extension of that for a synthetic electromagnetic
field. By adjusting the parameters of the atom-laser coupling and making
the energy of one of the hyperfine states much higher than the other two
in the rotating frame, one can effectively project out one hyperfine state
and the effective Hamiltonian can be re-arranged into the following form
under the basis of the remaining hyperfine states

$$
H_{\mathbf{k}} = \frac{\hbar^2}{2m} (\vec{k} + k_0 \vec{x} \sigma_z)^2 - \frac{\delta}{2} \sigma_z + \frac{\Omega}{2} \sigma_x \tag{11.10}
$$

$$
\Rightarrow \frac{\hbar^2}{2m} (\vec{k} + k_0 \vec{x} \sigma_x)^2 - \frac{\delta}{2} \sigma_x - \frac{\Omega}{2} \sigma_z, \tag{11.11}
$$

where we have rotated the spin basis in the second line so that the effective Hamiltonian takes the form of an equal mixture of Rashba $k_x \sigma_x + k_y \sigma_y$ and Dresselhaus $k_x \sigma_x - k_y \sigma_y$ spin-orbit coupling.

11.2.   Synthetic spin-orbit coupling

In the following, we will first focus on the effects of a more symmetric Rashba spin-orbit coupling, before briefly discussing the effects of the asymmetry of the NIST SOC on the system. The spin-orbit coupling has non-trivial effects on the system, many of which can be understood on the single-particle level. The single particle Hamiltonian under the Rashba spin-orbit coupling can be written as

$$H = \sum_{\mathbf{k},\sigma} \epsilon_{\mathbf{k}} a_{\mathbf{k},\sigma}^{\dagger} a_{\mathbf{k},\sigma} + \sum_{\mathbf{k}} \left[\alpha \left(k_x - i k_y \right) a_{\mathbf{k},\uparrow}^{\dagger} a_{\mathbf{k},\downarrow} + H.C. \right], \qquad (11.12)$$

where $a_{\mathbf{k},\sigma}^{\dagger}$ ($a_{\mathbf{k},\sigma}$) is the creation (annihilation) operator for the pseudo-spin $\sigma = \{\uparrow, \downarrow\}$ and α is the spin-orbit coupling strength. The pseudo-spin here is related to the adiabatic potential and is different from the hyperfine states in an atom. The Hamiltonian can be diagonalized by introducing the creation (annihilation) operator in the so-called helicity basis

$$a_{\mathbf{k},+} = \frac{1}{\sqrt{2}} \left(e^{i\varphi_{\mathbf{k}}} a_{\mathbf{k},\uparrow} + a_{\mathbf{k},\downarrow} \right), \qquad (11.13)$$

$$a_{\mathbf{k},-} = \frac{1}{\sqrt{2}} \left(e^{i\varphi_{\mathbf{k}}} a_{\mathbf{k},\uparrow} - a_{\mathbf{k},\downarrow} \right), \qquad (11.14)$$

where $\varphi = \arg(k_x, k_y)$. The diagonalized Hamiltonian

$$H = \sum_{\mathbf{k},\lambda=\pm} \xi_{\lambda} a_{\mathbf{k},\lambda}^{\dagger}, \qquad \xi_{\pm} = \epsilon_{\mathbf{k}} \pm \alpha k. \qquad (11.15)$$

The single particle spectra under the Rashba SOC are illustrated in Fig. 11.1, where it is clear that spin-orbit coupling breaks the inversion symmetry and splits the spin spectra into two helicity branches. Due to the symmetry of the Rashba spin-orbit coupling, points of the lowest energy form a ring in the momentum space and the ground state is infinitely degenerate in this case. Correspondingly, the density of states for the Rashba spin-orbit coupled spectra in three dimensions is a constant at the lowest energy. This peculiar single particle spectra and density of states naturally lead to interesting many-body phenomena [11, 16].

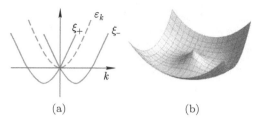

Figure 11.1. Illustration of the single particle spectra under the Rashba spin-orbit coupling (SOC). (a) The dashed curve represents the single particle spectra without spin-orbit coupling and the solid curves are the spectra for the helicity branches. (b) The three-dimensional view of the single particle spectrum of the lower helicity branch.

As the ground state degeneracy is infinite, there can no longer be a Bose–Einstein condensate at zero temperature, even in three dimensions, for non-interacting bosons. However, a weak interaction can induce a spontaneous symmetry breaking and the bosons will condense to either one or two opposite points on the degenerate ring in momentum space, depending on the interaction parameters. This leads to the so-called plane-wave phase and the striped phase in a uniform interacting Bose gas with the Rashba spin-orbit coupling [11, 12]. The spin-orbit coupling can also modify the collective excitation of the system. As spins are not of a Galilean covariant, the spin-orbit coupling will break the Galilean covariance of momentum as well. This can leave interesting signatures in the collective dipole oscillations, the critical velocity of superfluidity, and so on [7, 17]. The quantum and thermal fluctuations above the ground state are also changed, which eventually affect the condensation temperature [18–20]. Ever since the implementation of SOC in ultracold Bose gas, these have been under extensive theoretical and experimental study. Note that in experiments, only the asymmetrical NIST SOC has been realized, instead of the Rashba SOC. Under the NIST SOC, the degenerate ground state manifold consists of two points in the momentum space, in contrast to a ring under the Rashba SOC.

For a non-interacting Fermi gas at zero temperature, atoms occupy all the low-energy states up to the Fermi energy. As the Fermi energy increases with an increasing atom number, the topology of the Fermi surface changes while the system undergoes a Lifshitz transition. This has been experimentally observed in a non-interacting degenerate Fermi gas under the NIST scheme (see Fig. 11.3). On the other hand, it has been shown that for atoms with attractive s-wave interaction between the spin species and with

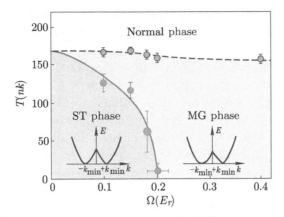

Figure 11.2.   Finite-temperature phase diagram of a BEC under the NIST SOC, where the boundaries of the striped phase (ST), the magnetized phase (MG) and the normal phase are determined experimentally [20].

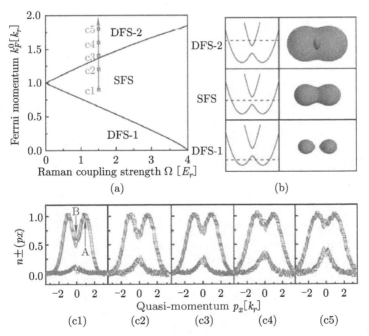

Figure 11.3.   Experimental observation of the Lifshitz transition for a non-interacting degenerate Fermi gas with spin-orbit coupling [5].

a Rashba spin-orbit coupling, the two-body bound state energy is enhanced due to the increased density of states at low energies [16, 21, 22]. In fact, a two-body bound state exists in three dimensions, even in the weak-coupling limit. This is similar to the case of a two-dimensional problem without spin-orbit coupling, which also has a constant density of states. The few-body features of the system inevitably affect the many-body properties. The BCS-BEC crossover problem of a three-dimensional two-component Fermi gas under a Rashba spin-orbit coupling has been studied in 2011 [16]. Basically, the spin-orbit coupling leads to an increased pairing order parameter and enhanced critical temperature throughout the crossover region [16, 22]. The pairing superfluidity of the system is particularly interesting, because singlet and triplet pairing states can coexist as spin-orbit coupling mixes up the spins.

In Fig. 11.1, the two helicity branches cross at the origin in the momentum space. Under an external Zeeman field along the z-direction, the helicity branches are coupled and a gap opens up at the origin. This kind of system has actually been studied before in the context of solid state materials, where it has been proposed that the coexistence of a Rashba spin-orbit coupling, an external Zeeman field and an s-wave pairing order can lead to the exotic topological superfluid phase [23]. The topological superfluid state features exotic edge states at the boundaries and can support vortex excitations with Majorana zero modes at the vortex core [8]. A natural motivation is the potential of studying the topological superfluid state in ultracold Fermi gases. During the past several years, a great amount of effort has been dedicated to the clarification of novel phases and phase transitions in a spin-orbit coupled, strongly interacting Fermi gas of ultracold atoms, where various forms of SOC in different dimensions can lead to a wealth of exotic superfluid phases [9, 10, 16, 21, 22, 24–36]. In three dimensions, two different topologically non-trivial superfluid phases with gapless excitations exist. In two dimensions, a topological superfluid phase which supports the Majorana zero mode at the core of vortex excitations can be stabilized. In a uniform system, these different phases will form phase-separated states with various first-order boundaries. In a trapping potential, which is more relevant to the experimental conditions, these different phases naturally phase separate in space, leading to shell structures similar to that of a polarized Fermi gas (see Fig. 11.4). The pairing physics under spin-orbit coupling is further enriched by the possibility of stabilizing pairing states with a non-zero center-of-mass momentum, the so-called FFLO states. This

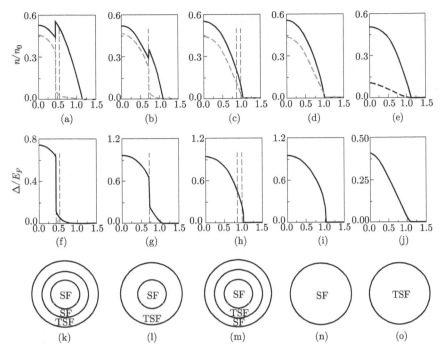

Figure 11.4.   Typical (a–e) density distribution, (f–j) order parameter and (k–o) phase structures of a two-dimensional Fermi gas with Rashba spin-orbit coupling and in a harmonic trapping potential [32]. The solid black (dashed) curves in the density subplots represent spin up (down) species. The thin dotted lines in the first two rows illustrate the TSF-SF or the SF-SF boundary.

can be achieved either by generating asymmetric single-particle dispersion [37] or by introducing spin-selective interactions [38]. Most interestingly, a topological FFLO phase — a novel topological pairing state — has been proposed to exist in these systems [39–41]. Owing to these latest developments, we will discuss the rich pairing physics under spin-orbit coupling in more detail in the next section.

11.3.   Exotic pairing states under spin-orbit coupling

Perhaps the most important exotic pairing state in a spin-orbit coupled Fermi system is the topological superfluid state in two dimensions, originally investigated in the context of semiconductor/superconductor heterstructures with a Rashba spin-orbit coupling, s-wave pairing order and an external Zeeman field. As has been shown in Refs. [9], [10] and [27], intra-branch pairing with p-wave symmetry in the helicity bases becomes possible as spin-orbit coupling breaks the inversion symmetry and mixes up different spin components in the helicity branches, in contrast to the conventional s-wave BCS pairing, where inter-species pairing is only allowed. The introduction of the external Zeeman field breaks the time-reversal symmetry and opens up a gap between different helicity branches. In the weak-coupling limit, when the Fermi surface of the system lies in this gap, the ground state of the system becomes a topological superfluid state with Majorana zero modes on the boundaries or at the core of vortex excitations. Importantly, these non-Abelian Majorana zero modes are protected by a bulk gap and are therefore useful for fault-tolerant topological quantum computations. The pairing scenarios for the BCS inter-species pairing and the spin-orbit coupling induced intra-branch pairing are illustrated in Fig. 11.5.

The new possibility of intra-branch pairing under spin-orbit coupling is a fundamentally new feature, which serves as the basis for various exotic pairing states in these systems. An important example is the dramatically enhanced Fulde–Ferrell pairing states under spin-orbit coupling and a Fermi surface asymmetry. This is illustrated in Fig. 11.5. Under an additional transverse Zeeman field, the Fermi surface is deformed, such that it no longer has inversion symmetry along the axis of the transverse field. In the weakly interacting limit, it is clear from the spectra that a simple BCS pairing state with zero center-of-mass momentum becomes energetically

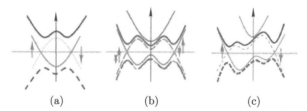

Figure 11.5.   Illustration of a pairing mechanism [37]. (a) Illustration of s-wave inter-species pairing in a conventional BCS pairing. (b) Spin-orbit coupling induced intra-branch pairing. (c) Pairing under spin-orbit coupling and Fermi surface asymmetry. In all cases, the thin (thick) solid curves are the dispersion spectra of (quasi) particles and the thin (thick) dashed curves are the dispersion spectra of (quasi) holes. The horizontal lines represent the Fermi surface.

unfavorable. This opens up the possibility of an exotic FFLO-like pairing state with finite center-of-mass momentum [37]. From the general argument above, we may further infer that such a pairing state is a natural result of the co-existence of spin-orbit coupling and Fermi surface asymmetry, and should generally exist in such systems, independent of the exact type of spin-orbit coupling and dimensionality. Indeed, these exotic pairing states have been reported to exist in various systems of different dimensions and with different forms of spin-orbit coupling [34, 35, 39–41].

To illustrate this, we first consider an experimentally relevant system, where NIST spin-orbit coupling is imposed on a two-dimensional Fermi gas with effective axial and transverse Zeeman fields. To understand the pairing states in such a system, the first necessary step is to characterize pairing states on the mean-field level, which should provide a qualitatively correct physics picture at zero temperature. For finite-temperature properties, one should resort to more involved calculations beyond the mean-field theory. The mean-field description of pairing states in a Fermi gas under spin-orbit coupling and a Zeeman field can be seen as a natural extension of the standard BCS theory. With spin-orbit coupling in the current case, the inversion symmetry of the system is broken and the effective Hamiltonian that follows from the mean-field approximation becomes

$$H_{\text{eff}} = \frac{1}{2} \sum_{\mathbf{k}} \begin{pmatrix} \lambda_k^+ & 0 & h & \Delta_Q \\ 0 & -\lambda_{Q-k}^+ & -\Delta_Q^* & -h \\ h & -\Delta_Q & \lambda_k^- & 0 \\ \Delta_Q^* & -h & 0 & -\lambda_{Q-k}^- \end{pmatrix} + \sum_{\mathbf{k}} \xi_{|\mathbf{Q}-\mathbf{k}|} - \frac{|\Delta_Q|^2}{U}, \quad (11.16)$$

where $\lambda_k^{\pm} = \xi_k \pm \alpha k_x \mp h_x$ and the order parameter $\Delta_Q = U \sum_{\mathbf{k}} \langle a_{\mathbf{Q}-\mathbf{k}\downarrow} a_{\mathbf{k}\uparrow} \rangle$. The Hamiltonian (11.16) has been written under the hyperfine-spin basis $\{a_{\mathbf{k}\uparrow}, a_{\mathbf{Q}-\mathbf{k}\uparrow}^{\dagger}, a_{\mathbf{k}\downarrow}, a_{\mathbf{Q}-\mathbf{k}\downarrow}^{\dagger}\}^T$. Experimentally, the spin-orbit coupling parameter α is related to the momentum transfer of the Raman process in the NIST scheme, and the effective Zeeman field h and h_x are proportional to the effective Rabi frequency and the two-photon detuning of the Raman lasers, respectively. The zero-temperature thermodynamic potential can be obtained by diagonalizing the effective Hamiltonian

$$\Omega = \sum_{\mathbf{k}} \xi_{|\mathbf{Q}-\mathbf{k}|} + \sum_{\mathbf{k},\nu} \theta(-E_{\mathbf{k},\nu}^{\eta}) E_{\mathbf{k},\nu}^{\eta} - \frac{|\Delta_Q|^2}{U}, \quad (11.17)$$

where the quasi-particle (hole) dispersion $E^{\eta}_{\mathbf{k},\nu}$ ($\nu = 1, 2$, $\eta = \pm$) are the eigenvalues of the matrix in Hamiltonian (11.16) and $\theta(x)$ is the Heaviside step function.

From the previous general analysis in the weak-coupling limit, the BCS pairing states with zero center-of-mass momentum would become unstable against a Fulde–Ferrell pairing state under the Fermi surface asymmetry. This implies an instability of the BCS state with a finite h_x. This point can be demonstrated by performing a small $\mathbf{Q}$ expansion around the local minimum in the thermodynamic potential landscape that corresponds to the BCS pairing state

$$\Omega(\Delta, Q_x) = \Omega_0(\Delta) + \Omega_1(\Delta)Q_x + \Omega_2(\Delta)Q_x^2 + \mathcal{O}(Q_x^3), \qquad (11.18)$$

where we have assumed $\mathbf{Q} = (Q_x, 0)$. It is then straightforward to numerically demonstrate that for $h_x = 0$, we have $\Omega_1 = 0$ and $\Omega_2 > 0$; while for $h_x \neq 0$, we have $\Omega_1 \neq 0$, which has an opposite sign to that of h_x. This is direct evidence that the BCS pairing state in the presence of a Fermi surface asymmetry and spin-orbit coupling becomes unstable against a Fulde–Ferrell state, with the center of mass momentum $\mathbf{Q}$ opposite to the direction of the transverse field h_x. A qualitative understanding of these Fulde–Ferrell states is that the combination of SOC and Fermi surface asymmetry shifts the local minima that corresponds to $Q = 0$ BCS pairing states onto the finite-$\mathbf{Q}$ plane. This is further reflected by the observation that the magnitude of the center-of-mass momentum Q decreases as the Fermi surface asymmetry becomes smaller, i.e., with increasing chemical potential μ or SOC strength α.

We show the mean-field phase diagram in Fig. 11.6. The most apparent feature of the phase diagram is the stabilization of FFLO$_x$ states over a large parameter region, with the center-of-mass momentum of the FFLO$_x$ states opposite to the direction of the transverse field. Furthermore, when the magnitude of Q_x becomes small enough, the nodal FFLO states cross a continuous phase boundary and become fully gapped. In the weak-coupling limit, the gapped FFLO (gFFLO) state can be understood as a pairing state within one helicity branch, where the Fermi surface deformation induced by a transverse field is accommodated by a finite center-of-mass momentum. Similar pairing states have been found in related systems with SOC and Fermi surface asymmetry [34, 35, 39–41]. Hence, these FFLO states are qualitatively different from the conventional FFLO states found in polarized Fermi gases.

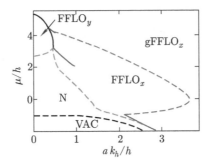

Figure 11.6.   Typical phase diagram for a two-dimensional Fermi gas under NIST SOC [37]. The solid curves are first-order boundaries and the dashed curves represent continuous phase boundaries.

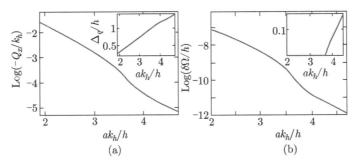

Figure 11.7.   (a) Evolution of the center-of-mass momentum and pairing order parameter (inset) for the ground state as the SOC strength increases. (b) Evolution of the difference in thermodynamic potential between the SF state and the FFLO_x state with increasing SOC strength, with $\delta\Omega = \Omega_{\mathrm{SF}} - \Omega_{\mathrm{FFLO}_x}$. The inset shows the evolution of the minimum excitation gap with increasing SOC strength. See Ref. [37]

From a theoretical perspective, a more interesting pairing state is perhaps the topological Fulde–Ferrell state, where the pairing state can simultaneously have non-zero center-of-mass momentum and topologically non-trivial properties. This exotic pairing phase can be understood as being derived from a typical topological superfluid phase in two dimensions, where the Rashba spin-orbit coupling, s-wave pairing order and an axial Zeeman field coexist. With the addition of another effective Zeeman field in the transverse direction, the Fermi surface becomes asymmetric, and according to the preceding analysis, the ground state of the system necessarily acquires a non-zero center-of-mass momentum. More importantly, the ground state would inherit all topological properties from the topological

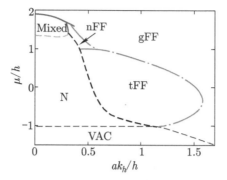

Figure 11.8.   Typical phase diagrams for the topological Fulde–Ferrell state on the α–μ plane [41]. The solid curves are first-order boundaries, while the dashed-dotted curves represent phase boundaries of continuous phase transitions. The dashed curves surrounding the normal region (N) are the threshold with $\Delta/h = 10^{-3}$, while the dotted curves are the boundary against the vacuum.

superfluid state, provided that the deformation of the Fermi surface should not be drastic enough to close the bulk gap. The phase diagram of a two-dimensional Fermi gas under a Rashba spin-orbit coupling and cross Zeeman fields is shown in Fig. 11.8.

Fermi surface asymmetry is not the only route toward a Fulde–Ferrell pairing in a spin-orbit coupled Fermi system. In fact, Fulde–Ferrell pairing can also be induced by spin-selective interaction together with spin-orbit coupling. This can take place, for example, in a three-component Fermi gas, where one fermion species is tuned close to a wide Feshbach resonance with one of the spin-species in a two-component Fermi gas under the NIST-type spin-orbit coupling [38]. The pairing mechanism is illustrated in Fig. 11.9, which should lead to a new type of Fulde–Ferrell pairing phase in the corresponding many-body system.

Besides the exotic pairing states discussed in this section, spin-orbit coupling has also been recently shown to give rise to novel universal trimer states, which feature exotic few-body correlations where even more interesting many-body phases may emerge [42, 43]. In all these cases, by modifying the single-particle dispersion, spin-orbit coupling proves to be a powerful tool of quantum control that when combined with the outstanding tunability of ultracold atomic gases, is playing an increasingly important role in fulfilling the potential of quantum simulation with ultracold atomic gases.

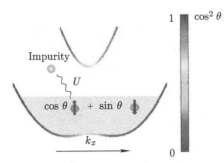

Figure 11.9. Schematics of the three-component Fermi-Fermi mixture [38]. The impurity atoms interact spin-selectively with a spin-orbit coupled two-component Fermi gas. The spin superpositions in both helicity branches are momentum-dependent, as characterized by θ_k. The pairing states naturally acquire a non-zero center-of-mass momentum in the system.

References

[1] Y.-J. Lin, R. L. Compton, A. R. Perry, W. D. Phillips, J. V. Porto and I. B. Spielman. *Phys. Rev. Lett.* **102**, 130401 (2009).

[2] Y.-J. Lin, R. L. Compton, K. Jimenez-Garcia, W. D. Phillips, J. V. Porto and I. B. Spielman. *Nature* **462**, 628 (2009).

[3] Y.-J. Lin, R. L. Compton, K. Jimenez-Garcia, W. D. Phillips, J. V. Porto and I. B. Spielman. *Nat. Phys.* **7**, 531 (2011).

[4] Y.-J. Lin, K. Jiménez-García and I. B. Spielman. *Nature (London)* **471**, 83 (2011).

[5] P. Wang, Z.-Q. Yu, Z. Fu, J. Miao, L. Huang, S. Chai, H. Zhai and J. Zhang. *Phys. Rev. Lett.* **109**, 095301 (2012).

[6] L. W. Cheuk, A. T. Sommer, Z. Hadzibabic, T. Yefsah, W. S. Bakr and M. W. Zwierlein. *Phys. Rev. Lett.* **109**, 095302 (2012).

[7] J.-Y. Zhang, S.-C. Ji, Z. Chen, L. Zhang, Z.-D. Du, B. Yan, G.-S. Pan, B. Zhao, Y. Deng, H. Zhai, S. Chen and J.-W. Pan. *Phys. Rev. Lett.* **109**, 115301 (2012).

[8] M. Z. Hasan and C. L. Kane. *Rev. Mod. Phys.* **82**, 3045 (2010).

[9] C. Zhang, S. Tewari, R. M. Lutchyn and S. Das Sarma. *Phys. Rev. Lett.* **101**, 160401 (2008).

[10] M. Sato, Y. Takahashi and S. Fujimoto. *Phys. Rev. Lett.* **103**, 020401 (2009).

[11] C. Wang, C. Gao, C. Jian and H. Zhai. *Phys. Rev. Lett.* **105**, 160403 (2010).

[12] C. Wu, I. Mondragon-Shem and X.-F. Zhou. *Chin. Phys. Lett.* **28**, 097102 (2011).

[13] J. Dalibard, F. Gerbier, G. Juzeliunas and P. Öhberg. *Rev. Mod. Phys.* **83**, 1523 (2011).

[14] X.-J. Liu, M. F. Borunda, X. Liu and J. Sinova. *Phys. Rev. Lett.* **102**, 046402 (2009).

[15] I. B. Spielman. *Phys. Rev. A* **79**, 063613 (2009).

[16] Z.-Q. Yu and H. Zhai. *Phys. Rev. Lett.* **107**, 195305 (2011).

[17] Q. Zhu, C. Zhang and B. Wu. *Euro. Phys. Lett.* **100**, 50003 (2012).

[18] T. Ozawa and G. Baym. *Phys. Rev. Lett.* **109**, 025301 (2012).

[19] T. Ozawa and G. Baym. *Phys. Rev. Lett.* **110**, 085304 (2013).

[20] S.-C. Ji, J.-Y. Zhang, L. Zhang, Z.-D. Du, W. Zheng, Y.-J. Deng, H. Zhai, S. Chen and J.-W. Pan. *Nature Phys.* **10**, 314 (2014).

[21] J. P. Vyasanakere, S. Zhang and V. B. Shenoy. *Phys. Rev. B* **84**, 014512 (2011).

[22] H. Hu, L. Jiang, X.-J. Liu and H. Pu. *Phys. Rev. Lett.* **107**, 195304 (2011).

[23] J. Alicea. *Rep. Prog. Phys.* **75**, 076501 (2012).

[24] M. Gong, S. Tewari and C. Zhang. *Phys. Rev. Lett.* **107**, 195303 (2011).

[25] M. Iskin and A. L. Subasi. *Phys. Rev. Lett.* **107**, 050402 (2011).

[26] W. Yi and G.-C. Guo. *Phys. Rev. A* **84**, 031608(R) (2011).

[27] L. Dell'Anna, G. Mazzarella and L. Salasnich. *Phys. Rev. A* **84**, 033633 (2011).

[28] M. Gong, G. Chen, S. Jia and C. Zhang. *Phys. Rev. Lett.* **109**, 105302 (2012).

[29] L. Han and C. A. R. Sá de Melo. *Phys. Rev. A* **85** 011606(R) (2012).

[30] R. Liao, Y. Yi-Xiang and W.-M. Liu. *Phys. Rev. Lett.* 108, 080406 (2012).

[31] L. He and X.-G. Huang. *Phys. Rev. Lett.* **108**, 145302 (2012).

[32] J. Zhou, W. Zhang and W. Yi. *Phys. Rev. A* **84**, 063603 (2011).

[33] X. Yang and S. Wan. *Phys. Rev. A* **85**, 023633 (2012).

[34] L. Dong, L. Jiang and H. Pu. *New J. Phys.* **15**, 075014 (2013).

[35] X.-F. Zhou, G.-C. Guo, W. Zhang and W. Yi. *Phys. Rev. A* **87**, 063606 (2013).

[36] X.-J. Liu, K. T. Law and T. K. Ng. *Phys. Rev. Lett.* **112**, 086401 (2014).

[37] F. Wu, G.-C. Guo, W. Zhang and W. Yi. *Phys. Rev. Lett.* **110**, 110401 (2013).

[38] L. Zhou, X. Cui and W. Yi. *Phys. Rev. Lett.* **112**, 195301 (2014).

[39] C. Chen. *Phys. Rev. Lett.* **111**, 235302 (2013).

[40] C. Qu, Z. Zheng, M. Gong, Y. Xu, L. Mao, X. Zou, G. Guo and C. Zhang. *Nat. Commun.* **4**, 2710 (2013).

[41] W. Zhang and W. Yi. *Nat. Commun.* **4**, 2711 (2013).

[42] Z.-Y. Shi, X. Cui and H. Zhai. *Phys. Rev. Lett.* **112**, 013201 (2014).

[43] X. Cui and W. Yi. *Phys. Rev. X*, 031026 (2014).

Part III

Quantum Simulation with Cold Atoms

12
Optical Lattice and Band Structure

Strongly correlated many-body physics is currently one of the central problems in physics. There are many important physical phenomena, such as high temperature superconductivity (high-T_c), concerned with strongly correlated many-body physics. Some seminal mechanics, such as the Hamiltonian, have been proposed to study these phenomena. For example, it is widely believed that the Fermi–Hubbard model plays a key role in high-T_c material. Unfortunately, the real physics underlying these phenomena is still unresolved; the main obstacle is that there is no efficient method to solve these Hamiltonian equations yet. For these strongly correlated many-body models, traditional analytic methods, such as the mean-field theory and perturbation theory, do not work. Powerful numerical methods, such as the Quantum Monte Carlo (QMC) and Density Matrix Renormalization Group (DMRG), also cannot give reasonable results for many of the relevant systems. The QMC method suffers from a sign problem for both frustrated systems and fermionic systems, and the DMRG method is limited to solving 1D and quasi 1D systems. To understand these key models and the strongly correlated physics, we need new ideas and technologies. The optical lattice, as a new platform, provides an opportunity to investigate these models. It serves as the bridge between physical theories and the physical phenomena of real materials.

In real materials, defects and disorders are unavoidable and make the mechanics under the physical phenomena more difficult to verify. The optical lattice is much cleaner than traditional materials. We can realize the desired "pure" models; furthermore, we can control the parameters in the

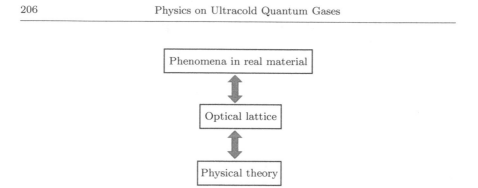

Figure 12.1. Optical lattices serve as the bridge between the physical phenomena in real materials and physical theories.

Hamiltonians to study the complete phase diagram of the models and monitor all dynamical processes between different phases. The controllability of optical lattices makes it possible to engineer a new Hamiltonian and investigate the physical regions that cannot be achieved through traditional methods. In this part of the book, we introduce this new platform and its applications in several difficult physical problems. Part III includes the following content: The construction of an optical lattice and the application of the platform to various problems, including the Bose–Hubbard (BH) model, the dynamical processes, the disordered systems and the spin systems.

12.1. Construction of optical lattices

Neutral atoms can be trapped in an optical dipole potential through the interaction with off-resonant light fields [1]. The strength of the trapping force is dependent on the intensity of the light field and the direction of the force depends on the detuning (red or blue). The periodicity of the light field induces periodic characteristics in the optical potential (neglecting the harmonic trap), which can play a similar role to a lattice in solid-state physics. Therefore, this periodic optical potential is named the optical lattice. The optical lattice is regarded as a good platform to simulate many-body physics in condensed matter systems. Before dicussing the use of an optical lattice, we first introduce how a lattice is constructed.

As previously described, a neutral atom in the far-detuned light field can feel a dipole potential [1] as

$$V_{dip} = \frac{3\pi c^2}{2\omega_0^3} \cdot \left(\frac{\Gamma}{\omega - \omega_0} + \frac{\Gamma}{\omega + \omega_0} \right) I(r), \qquad (12.1)$$

where c is the speed of light, ω_0 is the internal frequency of the atom, Γ is the lifetime of the atom level, ω is the frequency of the light field and $I(r)$ is the intensity of the light field at position r. Furthermore, if the condition $|\omega - \omega_0| \ll \omega_0$ is valid, the rotating-wave approximation [2] can be used to simplify the expression and obtain the formula

$$V_{dip}(r) = \frac{3\pi c^2}{2\omega_0^3} \frac{\Gamma}{\Delta} I(r), \qquad (12.2)$$

where $\Delta = \omega - \omega_0$ denotes the detuning of the light field. From this expression, we find that the dipole potential of a neutral atom in the light field is strongly dependent on the intensity and detuning. The topography of the potential is the same for the red- and blue-detuned fields; however, the atoms are trapped at different positions. If the light field is blue-detuned, that is, $\Delta > 0$, then the dipole potential is positive and the atom will prefer to remain at the position with the minimum light intensity. In contrast, if the light field is red-detuned, that is, $\Delta < 0$, then the atom will prefer to remain at the position with maximum light intensity.

Because the dipole potential is dependent on the light intensity, the optical lattice can be constructed and controlled by the periodic intensity of the field [3]. The simplest way to demonstrate construction of the periodic intensity is to consider the case of two counterpropagating plane waves [4]. Let there be two counterpropagating fields with the form

$$E_1(r,t) = E_0 Re[\epsilon exp(-i(\omega t - kz - \phi_1)] \qquad (12.3)$$

and

$$E_2(r,t) = E_0 Re[\epsilon exp(-i(\omega t + kz - \phi_2)], \qquad (12.4)$$

where E_0 is the amplitude of the fields, ω is the frequency of the fields, ϵ is the polarization of the fields, and both fields have the same polarization. If the two fields have different initial phases ϕ_1 and ϕ_2, then the superposition field of these two fields will be

$$E(r,t) = 2E_0 \cos(kz + (\phi_1 - \phi_2)/2) Re[\epsilon exp(-i(\omega t - (\phi_1 + \phi_2)/2], \qquad (12.5)$$

thus, the intensity of the fields is given by

$$I \approx 4E_0^2 \cos^2(kz + (\phi_1 - \phi_2)/2), \qquad (12.6)$$

which has the periodic term $\cos^2[kz + (\phi_1 - \phi_2)/2]$. Thus, this simple configuration has defined a lattice structure in the z-direction. However, this is not a real one-dimensional optical lattice because the atoms in the other two dimensions (x and y directions) are free. A real 1D optical lattice should be tightly confined on two dimensions (say x and y) and should have a larger gap between the ground state and the excitation state. Therefore, the system in the x and y directions will be frozen to the ground state and the low-energy character of this system is determined by the low-energy property of the z-direction interaction. As a result, this system can effectively be regarded as a one-dimensional system [5–9]. Using the same method, two-dimensional and three-dimensional optical lattices can be constructed by several counterpropagating fields (see Fig. 12.2).

In experiments, the light fields are not as simple as the ideal plane wave. We usually use laser beams to construct the optical lattice in experiment, and the characteristics of the laser beam have an important impact on the lattice. The intensity of a laser can be described by a Gaussian function [11]

$$I(r, z) = I_0 \left(\frac{w_0}{w(z)} \right)^2 exp(-2r^2/w(z)^2). \tag{12.7}$$

There are several parameters in this expression, which can be found in Fig. 12.3: I_0 denotes the intensity of the field at the center point of its waist (which is the narrowest point of the beam); w_0 denotes the waist size, with typical values of approximately $100\mu m$; r denotes the radial distance in the axial direction and z denotes the axial distance from the waist; $w(z) = w_0 \sqrt{1 + z^2/z_R^2}$ denotes the radius at which the intensity of the field will drop to $1/e^2$ of their axial values. The Rayleigh length z_R is defined as $\pi w_0^2/\lambda$ (λ is the wave length of the laser beam), which is a characterization of the asymptotic scaling of the beam width $w(z)$ with z. A typical Rayleigh length ranges from a millimeter to a centimeter.

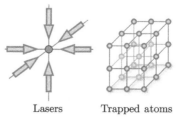

Lasers   Trapped atoms

Figure 12.2.   An illustration of the setup scheme and the resulting optical lattice in three dimensions.

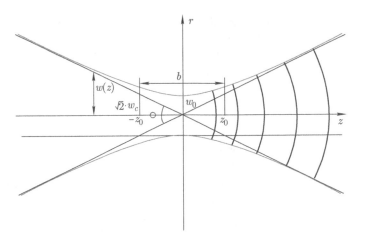

Figure 12.3. The intensity of the laser beam can be characterized as a Gaussian function. The notations are defined in the text.

Inserting the laser intensity formula Eq. (12.7) into Eq. (12.2), we can determine the expression of the dipole potential about the atom in the laser beam. If the atoms are located near the waist and the laser is red-tuned, the atoms will be attracted towards the waist center (where there is maximal intensity in the laser beam). We can approximate the dipole interaction near the waist of the laser beam as the harmonic potential by the Taylor expansion. That is,

$$V_{dip}(r, z) \approx -V_{trap}[1 - 2(r/w_0)^2 - (z/z_R)^2], \qquad (12.8)$$

where V_{trap} is linearly dependent on the power of the laser beam. The typical sizes of the waist and the Rayleigh length are much longer than the wavelength λ, which is the periodicity of the optical lattice. Thus, the periodic optical lattice is also a good approximation near the center point of the waist. However, a harmonic envelope should be added to the optical lattice away from the maximal intensity point, due to the Gaussian character of the laser beam. As a result, the potential of the optical lattice will be slightly modified with the Gaussian character of the laser beam,

$$V_{dip}(r, z) \simeq -V_0 exp(-2r^2/w(z)^2)sin^2(kz), \qquad (12.9)$$

where V_0 is the maximal depth of the lattice potential.

Generally, the atoms in the optical lattice are often trapped by an additional harmonic trap. In this situation, the harmonic trap can induce important physical effects that make detection and observation much more complicated in many-body systems. One such example is the Bose–Hubbard model in a harmonic trap, where the additional trap gives the phase diagram of the system a "wedding cake" configuration [12–14]. The coexistence of the different phases also makes observation much more difficult [15, 16].

12.2. Band structure

In this section, we neglect the harmonic potential in the optical lattice (the case of harmonic confinement can be found in Ref. [17]) for the purpose of simplicity. Therefore, the potential in an optical lattice is periodic and has translational symmetry. This is very similar to the situation in solid physics. It is well known that the eigenfunctions of a single particle in a potential with translational symmetry have a band structure [18]. That is,

Theorem 12.1. *Bloch theorem: The eigenfunction of such a periodic system can be written as the product of a plane wave envelope function $e^{ik \cdot r}$ and a periodic function $\mu_{n,k}(r)$ that is called the periodic Bloch function and has the same periodicity as the potential, i.e.,*

$$\phi_{n,\mathbf{k}}(\mathbf{r}) = e^{i\mathbf{k} \cdot \mathbf{r}} \mu_{n,\mathbf{k}}(\mathbf{r}).$$

The energies of the former eigenfunction are $E_n(\mathbf{k})$, with the relation $E_n(\mathbf{k}) = E_n(\mathbf{k} + \mathbf{K})$, where n is the index of the band, $\mathbf{k}$ is the wave vector and $\mathbf{K}$ is the periodicity of a reciprocal lattice vector. Within the same parameter n, the wave vector $\mathbf{k}$ may vary continuously and form the band structure. All Bloch functions form a complete orthonormal basis for the system, i.e., any function in this system can be expanded by Bloch functions.

This shows that Bloch functions have a plane-wave envelope. This characteristic makes Bloch functions more convenient to address systems that are almost free. That is, the periodic potential is small and the particles can move almost freely in that potential. In the opposite situation, that is, the periodic potential is strong and the particle cannot hop and is thus localized, the Bloch wave function is no longer the preferred basis, as the number of dominant Bloch functions will be very large. In this case, we prefer to use a localized basis of Wannier functions, $w_{n,\mathbf{R}}(\mathbf{r})$ (where n is also the band index, $\mathbf{R}$ is the lattice vector to denote the location of the function and $\mathbf{r}$ is the position of the particle).

The Wannier functions are also a complete orthonormal basis for the periodic system. Any of the operators in the system can be expanded by the Wannier functions. In particular, the Bloch functions can be expressed as a linear summation of the Wannier function, that is,

$$\phi_{n,\mathbf{k}}(\mathbf{r}) = \sum_{\mathbf{R}} w_{n,\mathbf{R}}(\mathbf{r}) e^{i\mathbf{k}\cdot\mathbf{R}}. \tag{12.10}$$

The Wannier functions are more suitable than the Bloch functions in the optical lattice situation. In practice, the optical lattice mainly works in the tight-binding regions and the Wannier functions have some useful characteristics that make them convenient to use. For example, at least for the lowest bands in the tight-binding model, the Wannier function is only dependent on the parameter $\mathbf{r} - \mathbf{R}$. Thus, the orthonormality of the functions can be expressed as

$$\int d^3 r w_n^*(\mathbf{r} - \mathbf{R}) w_{n'}(\mathbf{r} - \mathbf{R'}) = \delta_{n,n'} \delta_{\mathbf{R},\mathbf{R'}} \tag{12.11}$$

for different bands n and lattice vector $\mathbf{R}$.

For further use of the following, we expand a special operator that annihilates a particle at position $\mathbf{r}$, by the Wannier functions as

$$\hat{\phi}(\mathbf{r}) = \sum_{\mathbf{R},n} w_n(\mathbf{r} - \mathbf{R}) \hat{a}_{\mathbf{R},n}, \tag{12.12}$$

where $\hat{a}_{\mathbf{R},n}$ is the annihilation operator of the corresponding Wannier state. This expansion is frequently used in the tight-binding approximation. For convenience, a Hamiltonian is often expressed by the annihilation operator of $\hat{a}_{\mathbf{R},n}$ through the standard second quantization method. For example, the Hamiltonian of a particle moving in a periodic potential without any other interaction can be expressed with the operators $\hat{a}_{\mathbf{R},n}$ as

$$H_0 = \sum_{\mathbf{R},\mathbf{R'},n} t_n(\mathbf{R} - \mathbf{R'}) \hat{a}_{\mathbf{R},n}^\dagger \hat{a}_{\mathbf{R'},n}, \tag{12.13}$$

where $t_n(\mathbf{R} - \mathbf{R'})$ describes hopping within the same band n between the sites $\mathbf{R}$ and $\mathbf{R'}$. Generally, the Hamiltonian will be the sum of all the hopping between different sites $\mathbf{R}$ and $\mathbf{R'}$. Fortunately, in the tight-binding system, the hopping only happens between the nearest sites (other hopping is much less significant than the dominant hopping and can be neglected) and the Hamiltonian can be drastically simplified. Furthermore, the hopping matrix t_n is only determined by the overlap of the Wannier function with the same band n of the nearest sites.

The optical lattice in this section always works in the tight-binding region and we will address the Wannier functions only. It is worthwhile to describe the additional characteristics of the Wannier function and the hopping matrix [19]. In a one-dimensional situation, the Wannier function and the hopping matrix can analytically give their asymptotic behavior as [20]

$$w_n(r - R) \sim |r - R|^{-3/4} exp(-h_n|r - R|) t_n(R - R')$$
$$\sim |R - R'|^{-3/2} exp(-h_n|R - R'|), \qquad (12.14)$$

where $r - R$ denotes the distance of the particle around a site, $R - R'$ denotes the distance of the hopping and h_n is a constant for the position and increases with the band n. The analysis shows that the Wannier function is exponentially localized with the constant h_n. Additionally, the Wannier function does not uniformly converge to the local harmonic oscillator ground state of each well; $w_n(r - R)$ decays exponentially rather than in a Gaussian manner. However, in practice (where it is much more convenient to use a Gaussian function than the exact Wannier function in calculations), the local Gaussian ground state is a good approximation of the local Wannier function when the potential is sufficiently deep [11] (see Table 12.1).

Table 12.1.   The fidelity between the Wannier function, $|w\rangle$, and the local Gaussian ground function, $|\phi\rangle$, with different lattice depth. From Ref. [11]. Here, V_0 is the maximal strength of the dipole potential and E_r is the recoil energy of the atoms, which is a reasonable energy scale in the optical lattice. We often use the dimensionless parameter s to characterize the intensity of the optical lattice as sE_r. When the dimensionless parameter $s = V_0/E_r > 10$, the Gaussian ground state is a good approximation. That is, when we address the deep potential case, we use the Gaussian ground state as the local Wannier function, which is highly advantageous for the calculation.

V_0/E_r	$4J/E_r$	W/E_r	$J(2d)/J$	$|\langle w|\phi\rangle|^2$
3	0.444 109	0.451 894	0.101 075	0.971 9
5	0.263 069	0.264 211	0.051 641	0.983 6
10	0.076 730	0.076 747	0.011 846	0.993 8
15	0.026 075	0.026 076	0.003 459	0.996 4
20	0.009 965	0.009 965	0.001 184	0.997 5

References

[1] R. Grimm, M. Weidemuller and Y. B. Ovchinnikov. Advances in Atomic, *Molecular and Optical Physics* **42**, 95 (2000).

[2] L. Allen and J. H. Eberly. *Optical Resonance and Two-level Atoms*. Wiley, New York (1972).

[3] V. S. Letokhov, V. G. Minogin and B. D. Pavlik. *Opt. Commun.* **19**, 72 (1976).

[4] G. Grynberg and C. Robilliard. *Phys. Rep.* **355**, 335 (2001).

[5] M. Greiner, I. Bloch, M. O. Mandel, T. Hnsch and T. Esslinger. *Phys. Rev. Lett.* **87**, 160405 (2001).

[6] H. Moritz, T. Stferle, M. Khl and T. Esslinger. *Phys. Rev. Lett.* **91**, 250402 (2003).

[7] T. Kinoshita, T. Wenger and D. S. Weiss. *Science* **305**, 1125 (2004).

[8] B. Paredes, A. Widera, V. Murg, O. Mandel, S. Folling, J. I. Cirac, G. V. Shlyapnikov, T. W. Hansch and I. Bloch. *Nature* **429**, 277 (2004).

[9] B. L. Tolra, K. M. O'Hara, J. H. Huckans, W. D. Phillips, S. L. Rolston and J. V. Porto. *Phys. Rev. Lett.* **92**, 190401 (2004).

[10] D. Jaksch. Contemporary *Physics* **45**, 367 (2004).

[11] I. Bloch, J. Dalibard and W. Zwerger. *Rev. Mod. Phys.* **80**, 885 (2008).

[12] G. G. Batrouni, V. Rousseau, R. T. Scalettar, M. Rigol, A. Muramatsu, P. J. H. Denteneer and M. Troyer. *Phys. Rev. Lett.* **89**, 117203 (2002).

[13] M. Rigol, R. T. Scalettar, P. Sengupta and G. G. Batrouni. *Phys. Rev. B* **73**, 121103 (2006).

[14] C. Kollath, U. Schollwck, J. von Delft and W. Zwerger. *Phys. Rev. A* **69**, 031601 (2004).

[15] G. Campbell, J. Mun, M. Boyd, P. Medley, A. E. Leanhardt, L. G. Marcassa, D. E. Pritchard and W. Ketterle. *Science* **313**, 5787 (2006).

[16] S. Folling, A. Widera, T. Mller, F. Gerbier and I. Bloch. *Phys. Rev. Lett.* **97**, 060403 (2006).

[17] C. Hooley and J. Quintanilla. *Phys. Rev. Lett.* **93**, 080404 (2004).

[18] N. W. Ashcroft and N. D. Mermin. *Solid State Physics*. Holt, Rinehardt and Winston, New York (1976).

[19] W. Kohn. *Phys. Rev.* **115**, 809 (1959).

[20] L. He and D. Vanderbilt. *Phys. Rev. Lett.* **86**, 5341 (2001).

13
Simulation of the Bose–Hubbard Model

13.1. Introduction to the Bose–Hubbard model

The Bose–Hubbard (BH) model [1] is a famous and well-understood strongly correlated many-body model. It is conceptually very simple; however, it is complicated enough to support the rich phases. It has two very interesting phases: the Mott insulator (MI) phase and the superfluid (SF) phase. The MI [2] was first introduced by N. Mott to explain a new type of insulator that is incompatible with conventional band theories. This insulator is caused by the strong interaction between electrons, which is not considered in the conventional band theories. In Mott's theory [2], the strong interaction splits the original band into two (there is a gap between these two bands) and the lower band is filled with electrons. Therefore, the system is an insulator. The SF phase is a macroscopic quantum phase [3]. It behaves like a fluid with zero viscosity. In particular, Bose–Einstein condensation is in a superfluid state. These two phases play important roles in many-body physics; furthermore, there is a phase transition between them in the BH model [1]. This model is considered to be the standard model to test methods under investigation for the strongly correlated many-body phenomena. The Hamiltonian of the standard BH model is given by

$$H = -t \sum_{\langle i,j \rangle} (b_i^\dagger b_j + b_j^\dagger b_i) + \frac{U}{2} \sum_i \hat{n}_i(\hat{n}_i - 1) - \mu \sum_i \hat{n}_i, \qquad (13.1)$$

where t is the hopping amplification, U denotes the on-site repulsive interaction and μ is the chemical potential, and $\hat{n}_i = b_i^\dagger b_i$ is the number of
on the i-th site. There are two competing interactions in this model: the
hopping and the on-site repulsive interaction. The hopping aids the bosons
to move through the entire system, while the on-site repulsive interaction
induces the bosons to localize. When the hopping mechanism dominates,
the system should be in the SF phase with a long-range order. On the other
hand, when the localized mechanism dominates, the system will be in the
MI phase, which has integer bosons on each site. The competition of these
two mechanisms will drive a quantum phase transition.

The characteristics of the SF and MI phases can be summarized as
follows:

First, the MI phase occurs when $U \gg t$ in the BH model. The bosons
are in localized states with integer atoms occupying every site and the
corresponding state is

$$|\Psi_{\text{MI}}\rangle \propto \left(\prod_i b_i^\dagger \right) |0\rangle, \tag{13.2}$$

which has just one boson on each site. The MI phase is incompressible, that
is, $\partial \bar{n}/\partial \mu = 0$, which implies that the on-site density is unchanged with
increased chemical potential. The MI also has a finite gap. The incompress-
ibility can be used to define the MI phase. In addition, the correlation of
the MI phase decays exponentially, which can be used to distinguish it from
the SF phase.

Second, the SF phase is the phase when $U \ll t$. The bosons are
expanded to the whole system and the corresponding state is

$$|\Psi_{\text{SF}}\rangle \propto \frac{1}{\sqrt{N!}} \left(\sum_i b_i^\dagger \right)^N |0\rangle. \tag{13.3}$$

The SF phase is compressible and does not have a gap, and the correlation
of this phase decays algebraically. In addition, this phase can be determined
by its non-zero, off-diagonal long-range order.

Though this model is simple, it is difficult to be exactly solved with analytical methods. The mean-field method can determine the rough boundary
of the phases, which is the famous lobe phase diagram [1] (Fig. 13.1).

Generally speaking, if we fix the chemical potential μ/U, the system
will be in the MI phase when hopping is small and in a superfluid state

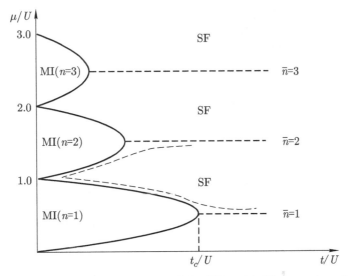

Figure 13.1. A schematic phase diagram of the BH model. There are two phases: the superfluid (SF) phase and the Mott insulator (MI) phase.

when μ is big enough. If we fix the hopping parameter t/U, the superfluid state and the MI phase will appear alternately Fig. 13.1.

However, the detailed boundary between the phases can only be calculated numerically. Different numerical methods may give a similar boundary shape but with slightly different details. These results of the boundary of the Mott insulating phase with $\langle n \rangle = 1$ in the one-dimensional case are collected in Fig. 13.2 [4], where "+" [5] and "×" [6] are determined by different sets of Quantum Monte Carlo (QMC) calculations; the solid line is an analysis of the 12th-order strong coupling expansions [7], the "filled circles" and "empty box" are the Density Matrix Renormalization Group (DMRG) results [8, 9] and the dashed line is the integer density's area. QMC is a very successful numerical method for boson systems; it is almost the standard method for that system. However, this method fails in frustrated systems and Fermi systems due to challenging sign problems. The DMRG method is another very precise numerical method in one-dimensional systems, but it is less powerful for two and higher dimension systems. For the one-dimensional boson system, these two methods are both valid and give similar phase diagrams. However, the boundary of the MI has some qualitative differences from the mean-field results. For example, it is not convex at the vertex point

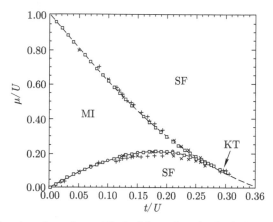

Figure 13.2. The phase boundary with the Mott phase in the BH model about the first lobe region determined by different numerical methods. The meaning of the different notations can be found in the text. From Ref. [4].

and there is a Kosterlitz–Thouless (KT) transition point. These details of the phase diagram cannot be found by the mean-field theory.

The MI phase [2] (in NiO) and the SF phase (in liquid Helium-4) have been observed in different materials. However, the two phases do not appear in the same material, i.e., there is no material where dominant physics has been described by the BH model. Thus, it is impossible to study the phase transition between the two phases, either by tuning the parameters of the hopping t or the interaction U in a traditional material. Fortunately, the optical lattice gives us a good platform to realize the BH model. All the parameters in the BH model can be controlled and tuned, so we can investigate the characteristics of the two phases and the quantum phase transition between them. More ambitiously, the dynamics near the phase transition can also be studied in this system. In the optical lattice, the intensity of the laser is the most convenient tuning parameter. In addition, the hopping between different sites is very sensitive to the intensity of the laser, as shown in the following derivation in the next section.

13.2.   Simulation of the Bose–Hubbard model in optical lattices

To realize the BH model, we use laser beams to construct a standard optical lattice, as introduced in the previous chapter. Then, we input the ultracold atoms that were already cooled by Bose–Einstein Condensation (BEC) into the lattice (the dipole potential is a bit shallow and it cannot confine the thermal atoms). The effective BH model can then be derived from the weakly interacting (with contact interaction) gas in the optical lattice under certain conditions [10]. The original Hamiltonian of the atoms in an optical lattice can be written as

$$H = \int d^3\mathbf{r}\psi^\dagger(\mathbf{r}) \left(-\frac{\hbar^2}{2m}\nabla^2 + V_0(\mathbf{r}) + V_T(\mathbf{r}) \right) \psi(\mathbf{r})$$

$$+ \frac{4\pi a_s\hbar^2}{2m} \int d^3\mathbf{r}\psi^\dagger(\mathbf{r})\psi^\dagger(\mathbf{r})\psi(\mathbf{r})\psi(\mathbf{r}), \qquad (13.4)$$

where $\psi(\mathbf{r})$ is a boson field operator for boson atoms in a given internal atom state, $V_0(\mathbf{r})$ is the optical lattice potential and $V_T(\mathbf{r})$ describes an additional external trapping potential, which is slowly varying compared to the optical potential $V_0(\mathbf{r})$. In a three-dimensional situation, the lattice potential is given by $V_0(\mathbf{r}) = \sum_{j=1}^{3} V_{j0} \sin^2(kr_j)$, where $k = 2\pi/\lambda$ and $a = \lambda/2$ is a lattice period. V_{j0} is the intensity of the field and it may depend on the site as a harmonic trap, as mentioned before. The interaction between bosons on the same site is a collision interaction that can be described and controlled by the S-wave scattering and the strength of the scattering can be described by the S-wave scattering length a_s. However, for Fermi atoms, S-wave scattering is prohibited by the Pauli principle and so the P-wave scattering may dominate.

We can only consider the system in the first band and then the boson operators $\psi(\mathbf{r})$ can be expanded by the Wannier states only in the first band. To make the single-band approximation valid, the system should satisfy the conditions that both the thermal energy and the mean interaction energy at a single site are much smaller than the gap between the first band and the first excited band. Thus, the atoms will stay in the first band and cannot be excited to the higher band. To achieve the effective Hamiltonian, we expand the boson field operator $\psi(\mathbf{r})$ by the Wannier function within the first band as

$$\psi(\mathbf{r}) = \sum_{\mathbf{R}} w(\mathbf{r} - \mathbf{R})\hat{b}_{\mathbf{R}}. \qquad (13.5)$$

We insert the expression of the boson operators into the original Hamiltonian. Considering the condition that the dipole potential is deep enough, i.e., since the Wannier functions effectively decay within a single lattice constant and the hopping happens only between the nearest neighbor sites. A standard BH model Hamiltonian can be achieved with the annihilation operators of the Wannier function.

$$H = -t \sum_{\langle i,j \rangle} (b_i^\dagger b_j + b_j^\dagger b_i) + \frac{U}{2} \sum_i \hat{n}_i(\hat{n}_i - 1) + \sum_i \epsilon_i \hat{n}_i. \tag{13.6}$$

The parameter U describes the strength of the on-site repulsion of atoms and it can be calculated by

$$U = 4\pi a_s \hbar^2 \int d^3\mathbf{r} |w(\mathbf{r})|^4 / m. \tag{13.7}$$

The parameter t that describes the hopping strength between the adjacent sites, which can be calculated by

$$t = \int d^3\mathbf{r} w(\mathbf{r} - \mathbf{R}_i) \left(-\frac{\hbar^2}{2m} \nabla^2 + V_0(\mathbf{r}) \right) w(\mathbf{r} - \mathbf{R}_{i+1}). \tag{13.8}$$

The other parameters ϵ_is are the energy offsets of each site, which play a similar role to the chemical potential in the Hamiltonian. They can be calculated by

$$\epsilon_i = \int d^3\mathbf{r} V_T(\mathbf{r}) |w(\mathbf{r} - \mathbf{R}_i)|^2. \tag{13.9}$$

It should be mentioned here that the parameters ϵ_is are site-dependent, which is different from the homogeneous chemical potential in the standard BH model.

From these expressions, it is clearly shown that the parameters t, U and ϵ_i are determined by the Wannier function and some of the system parameters, such as the intensity of the field V_0 and the scattering strength a_s. As mentioned in the characterization of Wannier functions, though the exact expressions of the Wannier functions are difficult to determine, the Wannier functions can be well approximated by the local Gaussian ground state when the optical potential is sufficiently deep, generally $V_0 \geq 10E_r$, where E_r is the recoil energy of the boson. Thus, the direct relation between the model parameters (t, U and ϵ_i) and the system parameters (V_0 and a_s) can be found. The hopping parameter t and repulsion parameter U in the

BH model can be expressed as the following

$$t = \frac{4}{\sqrt{\pi}} E_r \left(\frac{V_0}{E_r}\right)^{3/4} exp\left[-2\left(\frac{V_0}{E_r}\right)^{1/2}\right],$$

$$U = \sqrt{8/\pi} k a_s E_r \left(\frac{V_0}{E_r}\right)^{3/4}. \tag{13.10}$$

This shows that the effect of the intensity V_0 on the parameters U and t is substantially different, which can be clearly found by the ratio U/t

$$\frac{U}{t} \sim exp\left(2\sqrt{\frac{V_0}{E_r}}\right). \tag{13.11}$$

The parameter t is much more sensitive than the parameter U to the value of V_0. Their dependence on the parameter V_0 is shown in Fig. 13.3.

Thus, the ratio can be easily tuned by the laser intensity and drive the system into different phases.

We can now use this model to study the quantum phase transition between the MI and the SF phases. Before the experiment, we need to estimate the critical value of the experimental parameter V_0 at which the phase transition happens. The mean-field approximation can give a rough result of the critical value. From Fisher *et al.*'s [1], the critical point from the MI phase to the SF phase is given by $(U/tZ)_c \approx 5.8$ for $\bar{n} = 1$ and $(U/tZ)_c \approx 4\bar{n}$ for $\bar{n} \gg 1$. $\bar{n}$ denotes the integer of on-site bosons in the MI phase and $Z = 2d$ (where d is the dimension) is the number of nearest neighbors. For convenience, we always tune the parameters to make the on-site atoms close to 1 in experiments. In this situation, much more precise results for the critical values [11] can be found with some numerical

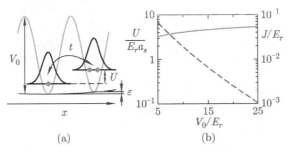

(a)   (b)

Figure 13.3.   The dependence of the parameters, t and U, in the BH model and the parameter, V_0, corresponding to the intensity of the laser. From Ref. [10].

calculations: $(U/t)_c = 29.36$ for a simple cubic lattice by QMC methods and $(U/t)_c = 3.37$ for a one dimension lattice by the DMRG method [8, 9]. Therefore, when we tune the intensity of the laser beam to make the parameter U/t cross the critical value, the MI-SF phase transition will be observed.

So far, we have neglected the effect of the harmonic trap. Now, we add this effect to the BH model. It fills the same role as the chemical potential. In fact, the standard BH model has a homogeneous chemical potential (the chemical potential is the same for every site), while the BH model with a harmonic trap has an inhomogeneous chemical potential ($\mu_{\mathbf{R}} = \mu(0) - \epsilon_{\mathbf{R}}$ with $\epsilon_{\mathbf{R}} = 0$ at the center of the trap). The inhomogeneity will cause the phases of the BH model to have a "wedding cake" configuration (see Fig. 13.4) [12–16].

In Fig. 13.4, we suppose the center parameter falls into the MI phase with $\bar{n} = 2$. The inhomogeneity will make several different phases coexist in the trap, which makes detection and observation more difficult. General observation technology, such as the TOF, could be affected by the mixture of the phases. To overcome this difficulty, new detection technologies have been developed to observe the system site by site [15, 16].

With the former method, the BH model has been experimentally simulated in an optical lattice [18]. Through tuning the depth of the optical lattice V_0 and by detecting the distribution of the momentum measurements

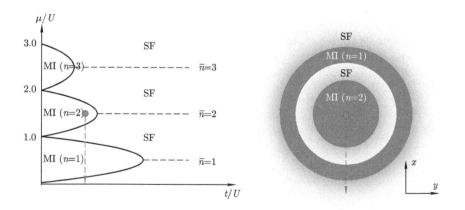

Figure 13.4.   The "wedding cake" configuration of the phases of the BH model in a harmonic trap.

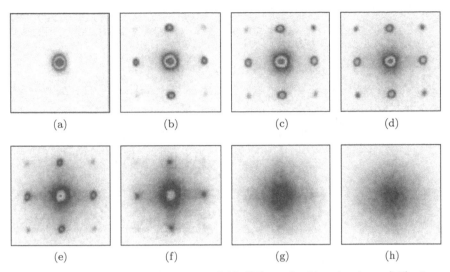

(a)　　　　　　(b)　　　　　　(c)　　　　　　(d)

(e)　　　　　　(f)　　　　　　(g)　　　　　　(h)

Figure 13.5. Observations of the superfluid (SF) to the Mott insulator (MI) phase transition in an optical lattice. From Ref. [18].

by the TOF, the following experimental results have been obtained to show the SF-MI phase transition [18–21].

The momentum distributions of the atoms with different V_0 are shown in Fig. 13.5(a) to 13.5(h) and their values are: $0E_r$, $3E_r$, $7E_r$, $10E_r$, $13E_r$, $14E_r$, $16E_r$ and $20E_r$. At the beginning, the system is in the SF phase. When the potential height reaches approximately $13E_r$ (**e**), the incoherent atoms begin to increase until the potential height is approximately $22E_r$ and the coherent atoms disappear entirely. This transition process has been clearly demonstrated by tuning the parameter V_0.

After the first realization of the BH model and observation of the MI-SF transition in an optical lattice, the BH model in an optical lattice has been viewed as a toolbox to simulate a variety of many-body models. Simulation in an optical lattice opens up a novel and promising way in understanding many long-standing problems that have been intractable, such as many-body systems in condensed matter.

References

[1] M. P. A. Fisher, P. B. Weichman, G. Grinstein and D. S. Fisher. *Phys. Rev. B* **40**, 546 (1989).

[2] N. F. Mott. *Metal-Insulator Transitions*. Taylor & Francis, London (1990).

[3]   A. M. Gunault. *Basic Superfluids.* Taylor & Francis, London (2003).
[4]   M. A. Cazalilla, R. Citro, T. Giamarchi, E. Orignac and M. Rigol. *Rev. Mod. Phys.* **83**, 1405 (2011).
[5]   G. G. Batrouni, R. T. Scalettar and G. T. Zimanyi. *Phys. Rev. Lett.* **65**, 1765 (1990).
[6]   V. A. Kashurnikov, A. V. Krasavin and B. V. Svistunov. *JETP Lett.* **64**, 99 (1996).
[7]   N. Elstner and H. Monien. *Phys. Rev. B* **59**, 12184 (1999).
[8]   T. D. Kuhner and H. Monien. *Phys. Rev. B* **58**, 14741(R) (1998).
[9]   T. D. Kuhner, S. R. White and H. Monien. *Phys. Rev. B* **61**, 12474 (2000).
[10] D. Jaksch, C. Bruder, J. I. Cirac, C. W. Gardiner and P. Zoller. *Phys. Rev. Lett.* **81**, 3108 (1998).
[11] B. Capogrosso-Sansone, N. V. Prokofev and B. V. Svistunov. *Phys. Rev. B* **75**, 134302 (2007).
[12] G. G. Batrouni, V. Rousseau, R. T. Scalettar, M. Rigol, A. Muramatsu, P. J. H. Denteneer and M. Troyer. *Phys. Rev. Lett.* **89**, 117203 (2002).
[13] M. Rigol, R. T. Scalettar, P. Sengupta and G. G. Batrouni. *Phys. Rev. B* **73**, 121103 (2006).
[14] C. Kollath, U. Schollwck, J. von Delft and W. Zwerger. *Phys. Rev. A* **69**, 031601 (2004).
[15] G. Campbell, J. Mun, M. Boyd, P. Medley, A. E. Leanhardt, L. G. Marcassa, D. E. Pritchard and W. Ketterle. *Science* **313**, 5787 (2006).
[16] S. Folling, A. Widera, T. Mller, F. Gerbier and I. Bloch. *Phys. Rev. Lett.* **97**, 060403 (2006).
[17] I. Bloch, J. Dalibard and W. Zwerger. *Rev. Mod. Phys.* **80**, 885 (2008).
[18] M. Greiner, M. O. Mandel, T. Esslinger, T. Hnsch and I. Bloch. *Nature* **415**, 39 (2002).
[19] T. Stferle, H. Moritz, C. Schori, M. Koehl and T. Esslinger. *Phys. Rev. Lett.* **92**, 130403 (2004).
[20] I. B. Spielman, W. D. Phillips and J. V. Porto. *Phys. Rev. Lett.* **98**, 080404 (2007).
[21] I. B. Spielman, W. D. Phillips and J. V. Porto. *Phys. Rev. Lett.* **100**, 120402 (2008).

14
Dynamical Process

Optical lattices provide a platform to continuously tune the system parameters. This characteristic makes it very convenient to investigate the dynamic process that cannot be achieved in conventional material. We will introduce two related processes that have been experimentally studied in the optical lattice: the quench dynamics near the phase transition and thermalization processes.

14.1. Quench dynamics in the Bose–Hubbard model

In the previous chapter, the Bose–Hubbard (BH) model was realized in an optical lattice [1]. When the system is in the weak interaction regime, that is, U is small compared to t, the system will be in a superfluid (SF) state. If we suddenly change the parameter V_0 to the MI regime, the system will evolve with the new Hamiltonian. Because the sudden change and the initial SF state is not an eigenvector of the new Hamiltonian, the state cannot adiabatically evolve to an equilibrium state. This process is a dynamical process with the non-equilibrium initial SF state. Theoretically, we can investigate this process by expanding the SF state on the eigenvectors of the new Hamiltonian whose ground state is a MI state.

$$|\Psi\rangle_{\text{SF}} = \sum_j a_j |\varphi_j\rangle_{\text{MI}}. \tag{14.1}$$

The real-time evolution is just a modification of the phases before the MI basis,

$$e^{-iHt}|\Psi\rangle_{\text{SF}} = \sum_j a_j e^{-iE_j t}|\varphi_j\rangle_{\text{MI}}, \qquad (14.2)$$

where E_j is the eigenvalue of the Hamiltonian H corresponding to eigenvector $|\varphi_j\rangle_{MI}$.

However, there are many eigenvectors, and we cannot analytically determine all the phases and sum the items to produce an analytical result. Generally, the eigenvectors are not limited to the first band of the system (which we previously used as a condition in the optical lattice to obtain the BH model). However, when we only quench the system near the critical point, the approximation is also valid due to the similarity of their energies, which constrain the states involved in this process. The energy is converved during the entire evolution process.

The dynamics of the quench process can only be theoretically studied by some numerical methods. The calculation also suffers from a rapidly increasing error with increasing quench time t (the error will accumulate during the evolution). When the time t is sufficiently large (the concrete time depends on the concrete model), the system error will go beyond the threshold. The effective numerical methods to exploit the dynamical process are the time Density Matrix Renormalization Group tDMRG1, tDMRG2, tDMRG3 [2–4] method and the time evolution block decimation (TEBD) algorithms in one dimension [5, 6]. For higher dimensions, we can only use the real-time evolution algorithms based on the tensor network state (TNS) [7, 8], which is an effective representation of the many-body states. The 1D tDMRG and TEBD algorithms are both based on the Matrix Product State (MPS) [9–11], which is a special TNS, such that in all the time evolution algorithms we must express the many-body state of the system as a TNS. For example, the one-dimensional MPS can be represented as in Fig. 14.1 [5, 6].

Then, we use the Suzuki–Trotter expansion [12, 13] to expand the nonlocal time evolution operator of the entire system into a local time evolution operator with controllable errors. The errors can be controlled as $\Delta \propto \delta t^\alpha$, where δt is the step of the evolution time and α is the order of the expansion. For example, the second-order Suzuki–Trotter

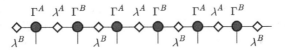

Figure 14.1.   The MPS representation of a one-dimensional system, where Γ^A and Γ^B are tensors that represent the projection from the virtual particle to the physical index. λ^A and λ^B are diagonal matrices whose diagonal elements represent the Schmidt coefficients of the many-body state if they divide the whole system from their location. Any two particle states can be represented as $|\Psi\rangle_{C,B} = \sum_i a_i |i\rangle_C |i\rangle_D$ under the local unitary transformation. The coefficients a_i are called the Schmidt coefficients.

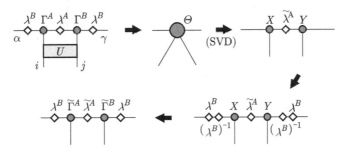

Figure 14.2.   A singular value decomposition process.

expansion is

$$e^{-i2\delta t(H_1 + H_2 + \cdots + H_N)} = e^{-i\delta t H_1} e^{-i\delta t H_2} \cdots e^{-i\delta t H_N}$$

$$\times e^{-i\delta t H_N} e^{-i\delta t H_{N-1}} \cdots e^{-i\delta t H_1}, \qquad (14.3)$$

where $H_i (i = 1, 2, \cdots, N)$ is the local Hamiltonian, only including the local operators.

After the local evolution, the inner dimension of the tensor will increase exponentially with the number of the local evolutions (or with the evolution time) and quickly exceed our computation capability. We can use some approximations, such as reducing the inner dimension (the number of the Schmidt coefficients) of the tensor to some fixed number after each local evolution, to avoid the exponential increase. This process can be completed by the Singular Value Decomposition (SVD), as in Fig. 14.2 [14].

In a one-dimensional case, the fixed dimension of the tensor is closely related to the entanglement of the system, which is characterized by the block entropy [15]. When the entanglement is small, the many-body state can be very well approximated by the tensor with a small inner dimension. However, if the entanglement is large, the inner dimension will be large. Unfortunately, the entanglement of the system will increase with the evolution. Because the dimension is fixed, the approximation will eventually fail. In calculations, we use the energy as an index to check the validity of the approximation because it is conserved during the evolution. Using the tDMRG methods [16], the SF-MI quenching process can be numerically simulated (see Fig. 14.3).

The key common characteristic of the quenching of the superfluid state is that the coherence of the initial superfluid state will be destroyed and be revived after some time. Because the BH model has been realized in the optical lattice, this quenching process can also be observed. The quench process [17] and the revival of the coherence are indeed observed during the experiment and the results in Fig. 14.4 are given by TOF imaging.

The destruction and revival processes can be found more clearly from the ratio of the coherence number and the total number of atoms in the optical lattice (see Fig. 14.5).

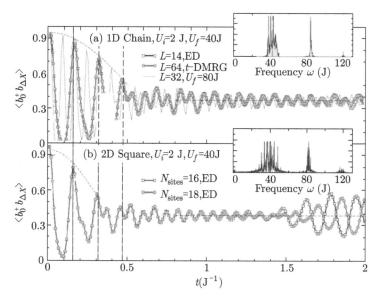

Figure 14.3.   A comparison of the results from exact diagonalization (ED) and tDMRG for a numerical SF-MI quenching process in one and two dimensions. For a very small system, ED is an exact method and can be used for comparison with the other method. From Ref. [16].

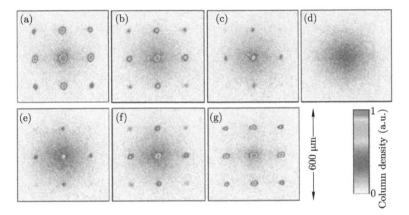

Figure 14.4. The experimental results of the SF-MI quenching process by TOF. At the beginning, the state is superfluid (a), then the state evolves to a Mott-like state (d) and finally, the system returns to the superfluid state (g). From Ref. [17].

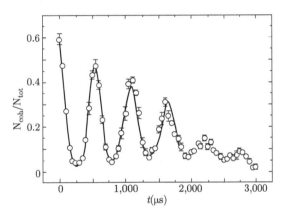

Figure 14.5. The experiment of SF-MI quench to observe the revival of the coherent. From Ref. [17].

14.2. Thermalization in an optical lattice

The quantum quench from the SF to the MI state is a special case of the thermalization process. In general statistical physics, the final state of a closed system is a thermal state that is only dependent on the temperature (determined by the initial energy) and that is independent of the details of the initial state [18]. With this assertion, the form of the final state of the SF-MI quenching process is a thermal state determined by the Hamiltonian in the MI region.

Generally, a thermal state is described by the Boltzmann distribution,

$$\rho_{th} = Z \sum_i e^{-E_i/KT} |\psi_i\rangle\langle\psi_i|, \qquad (14.4)$$

where Z is the normalization factor, E_i are the eigenvalues of the system, K is the Boltzmann constant, T is the temperature and $|\psi_i\rangle$ is the corresponding eigenvector.

Per the general thermalization theory, the initial SF state only affects the quenching process; it does not affect the final state. However, the general knowledge of the thermalization of a closed system is not always correct, especially for integrable systems [19, 20], in quantum systems. In fact, the thermalization problem can be divided into two separate but related problems:

- Is there a steady-state of the thermalization of a closed system?
- Is this steady-state a thermal state (or how would you describe the steady-state)?

For the first question, there is a closely related question: how to judge whether the state is steady? Generally, we determine that a state is steady according to the stability of some physical quantities, such as correlations. It is worth noting that the stability of some physical quantities do not directly mean that the steady-state is a constant state. If the steady-state does exist, how is it described? Ideally, the steady-state can only be determined by some physical quantities in the initial state and the Hamiltonian of a closed system, and does not depend on the details of the thermalization process.

The integrable systems are good examples to show that these two problems are separate. The thermalization of the integrable system is actually a steady-state. However, it cannot be described by a Boltzmann distribution but rather by another distribution known as the generalized Gibbs ensemble [21, 22].

The term integrable is used in several different situations, which are defined in different ways. Some of them are listed as follows. For a system to be integrable, it must satisfy one of the following conditions [22–24]:

- The quantum many-body system is exactly solvable.
- There are n (n is the number of degrees of freedom in the system) independent operators that are commutable with each other (the eigenvectors of the system are the common eigenvectors of these operators).

- The system is integrable by the Bethe ansatz.
- The system exhibits non-diffractive scattering [23, 24].

The Tonks–Girardeau (TG) gas [25] is a typical integrable system. This model has been realized in an optical lattice and the thermalization process of this system has also been observed. However, experimental results show that the TG gas cannot be thermalized. This is an important example that shows an integrable system cannot be thermalized.

The TG gas is a special case of the Lieb–Liniger gas [26]. The Hamiltonian of the one-dimensional Lieb–Liniger model is

$$H = -\frac{\hbar^2}{2m} \left(\sum_{i=1}^{N} \frac{\partial^2}{\partial x_i^2} + 2c \sum_{i<j=1}^{N} \delta(x_i - x_j) \right), \qquad (14.5)$$

when $c = 0$, this is the free bosons model. When $c \to \infty$, which is the hard-core or TG limit, this model becomes the TG gas model. The hard-core limit means that there is infinitely strong contact repulsion between bosons at the same site, which implies that any many-body wave function of the TG gas must vanish whenever two atoms meet at the same point. This key condition can be satisfied by the following form

$$\psi^B(x_1, x_2, \ldots, x_N) = S(x_1, x_2, \ldots, x_N)\psi^F(x_1, x_2, \ldots, x_N), \qquad (14.6)$$

where ψ^F is a wave function of a Fermi system, which naturally satisfies the former constraint with the Pauli principle. The function $S(x_1, x_2, \ldots, x_N) = \prod_{i>j=1}^{N} \text{sign}(x_i - x_j)$ is a sign function, which modifies the sign of the right-hand side wave function to satisfy the boson statistics.

Suppose that the system has an additional periodic boundary condition (period L), then the ground state of the free Fermi gas $\psi_0^F(x_1, x_2, \ldots, x_N)$ can be expressed by the well-known Slater determinant of the plane wave $e^{2\pi p x / L}$ (p is integer)

$$\psi_0^F(x_1, x_2, \ldots, x_N) = C e^{-i(n-1)\pi \sum_j x_j / L} \begin{pmatrix} 1 & Z_1 & Z_1^2 & \cdots & Z_1^{n-1} \\ 1 & Z_2 & Z_2^2 & \cdots & Z_2^{n-1} \\ \vdots & \cdots & \ddots & \cdots & \vdots \\ 1 & Z_n & Z_n^2 & \cdots & Z_n^{n-1}, \end{pmatrix}, \qquad (14.7)$$

where $Z_k = e^{i2\pi x_k / L}$ and C is the normalization factor. The former determinant is well-studied and can be calculated directly to give the explicit form

$$\psi_0^F(x_1, x_2, \ldots, x_N) = Ce^{-i(n-1)\pi \sum_j X_j/L} \prod_{j>l}(e^{i2\pi x_j/L} - e^{i2\pi x_l/L}). \quad (14.8)$$

Thus, the ground state of the Boson gas is given by

$$\psi_0^B(x_1, x_2, \ldots, x_N) = |\psi_0^F(x_1, x_2, \ldots, x_N)| \propto \prod_{i<j} sin\frac{\pi}{L}|x_i - x_j|. \quad (14.9)$$

So far, the TG gas has been solved exactly. In addition, the energy of this system can also be determined. For example, the ground energy of the TG gas is $E = \frac{\hbar^2(\pi\rho_0)^2}{6\pi m}$, where $\rho_0 = N/L$ is the mean particle density.

The former solution is focused on the wave function of the TG gas. This model can also be solved in another way, by finding sufficient integrals of motion. We first express the model in the second quantum quantized form

$$\hat{H} = -J\sum_{i=1}^{L}(\hat{b}_i^\dagger \hat{b}_{i+1} + H.C.), \quad (14.10)$$

where $\hat{b}_i$ is a boson and satisfies $(\hat{b}_i)^2 = (\hat{b}_i^\dagger)^2 = 0$ (by the Tonks–Girardeau limit). Generally, a boson operator can be transformed into a Fermi operator by the Jordan–Wigner transformation.

$$\hat{b}_i^\dagger = \hat{c}_i^\dagger \prod_{i'=1}^{i-1} e^{-i\pi \hat{c}_{i'}^\dagger \hat{c}_{i'}},$$

$$\hat{b}_i = \prod_{i'=1}^{i-1} e^{-i\pi \hat{c}_{i'}^\dagger \hat{c}_{i'}} \hat{c}_i. \quad (14.11)$$

With this transformation, the new fermion system is a non-interacting system, that is

$$\hat{H} = -J\sum_{i=1}^{L}(\hat{c}_i^\dagger \hat{c}_{i+1} + H.C.), \quad (14.12)$$

where $\hat{c}_i$ is a fermion operator.

For this free system, we can find sufficient integrals of motion [23, 24] to make it integrable. These integrals are

$$\hat{I}_k = \hat{f}^F(k) = \frac{1}{L} \sum_{i=1}^{L} \sum_{i'=1}^{L} \sigma_{i-i'}(\hat{N}) e^{-i2\pi k(i-i')/L} \hat{c}_{i'}^{\dagger} \hat{c}_i, \qquad (14.13)$$

where $\sigma_{\Delta_i(\hat{N})} = 1$ for odd N and $\sigma_{\Delta_i(\hat{N})} = e^{-i\pi\Delta_i/L}$ for even N. The number of the integrals is the same as the number of sites. Thus, the TG gas is an integrable system by the listed criteria.

The optical lattice is an effective platform to study the thermalization process. It is well isolated from the environment and can be viewed as a closed system. It has been used to realize the TG gas and to study the thermalization process [27, 28]. The TG gas is a typical one-dimensional system. We need to construct a real one-dimensional optical lattice in the experiment. Two orthogonal blue-detuned standing waves with a minimal frequency difference are used to form a two-dimensional optical lattice that is sufficiently strong to provide tight transverse confinement. That is, the lowest transverse excitation, denoted by $\hbar\omega_r$, where $\omega_r/2\pi$ is the transverse oscillation frequency, far exceeds all other energies and the system will stay in the ground state of the transverse directions. This two-dimensional optical lattice creates an array of 1D tubes (see Fig. 14.6) and the tunneling among the tubes can be negligible. Under these conditions, the system in each tube is strictly a one-dimensional system.

Though the 1D boson system is reached, it is difficult to reach the TG region, which needs a very strong interaction. The TG region can be characterized by the ratio $\gamma = \epsilon_{Int}/\epsilon_{Kin}$, where ϵ_{Int} and ϵ_{Kin} are the

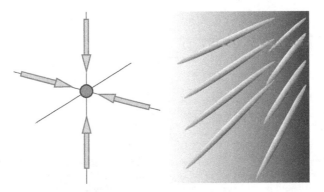

Figure 14.6.   Illustration of the laser setup to construct one-dimensional tubes.

average of the interaction and kinetic energy of a particle, respectively. For a very large γ value, the 1D system will be in the TG region. There are several ways to increase γ:

First, by increasing the effective interaction strength, that is, increasing ϵ_{Int}. Second, by decreasing the density of the quantum gas. ϵ_{Int} and ϵ_{Kin} depend differently on the density n; ϵ_{Kin} decreases at a rate of n^2, while ϵ_{Int} decreases with n. The former is a much faster rate than the latter. Thus, in the homogeneous gas, the parameter γ can be expressed as

$$\gamma = mg/\hbar^2 n, \qquad (14.14)$$

where m is the mass of a single atom, g is the interaction strength and n is the atomic density. Third, by adding an additional red-tuned optical lattice along the 1D tubes. The optical lattice will increase the effective mass of the atoms and increase the parameter γ to reach the TG region.

With these technologies, the TG gas can be achieved with an array of tubes, a combination of two blue-tuned 2D optical lattices and a crossed dipole trap with $\gamma \gg 1$. The typical characteristic of the TG gas is fermionization. This is readily observed in experiments by several research groups [27, 28] (see Fig. 14.7).

After the realization of the TG gas, we used this system to investigate the thermalization of the integrable system [30]. We prepared the initial

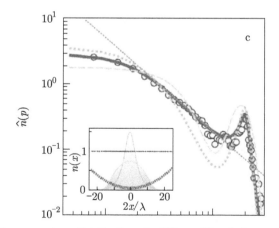

Figure 14.7.   The momentum distribution of a TG gas. The circles represent the experimental data, which are averaged over all the tubes; the black solid line is the best fit of these data and the dashed line represents the ideal Fermi gas. For comparison with the TG gas, the dotted line represents the ideal Bose gas. From Ref. [27].

state with a non-equilibrium momentum distribution by sending two pulses along the tubes to deplete the zero momentum state and transfer atoms to the $\pm 2\hbar k$ peaks (k is the wave vector of the optical light along the tubes). This state will evolve under the integrable TG gas Hamiltonian in each tube. In this experiment, the expanded momentum distributions, $f(P)$, are measured by the absorption image. The experimental results of the momentum distribution with different coupling strengths and evolution times are shown in Figs. 14.8 and 14.9.

In Fig. 14.8, the shapes of curves 2 and 3 are similar and curve 3 is lower than curve 2. This change is due to known loss and heating. If we consider the effect of the loss and heating and rescale the momentum distribution, we can then obtain the typical distribution, as in Fig. 14.9.

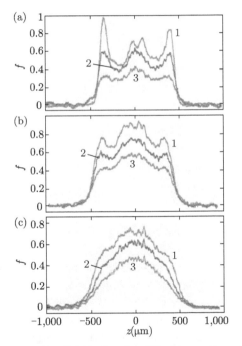

Figure 14.8.   The momentum distribution for different couplings. From Ref. [30]. The coupling strength in (b) (approximately 1) is larger than the coupling strength in (c) (approximately 0.62). Curve 1 is the momentum distribution, which is time-averaged over the first cycle. Curves 2 and 3 are the momentum distributions at time 15τ and 40τ (where τ is the period of axial oscillation), respectively. The experiment shows that the shape of the distribution will be stable after 10τ evolution. That is, the system tends to a steady-state.

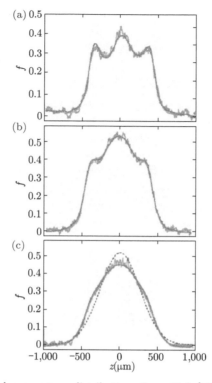

Figure 14.9.   Rescaled momentum distribution. From Ref. [30]. The blue and green curves are momentum distributions, $f(P)$, rescaled to consider the loss and the known heating. The red curve is the actual distribution.

If the system can be thermalized, the momentum distribution of the thermalized state is a Gaussian distribution due to the average of the many 1D TG gas tubes. The dashed line in the figure represents the Gaussian with the same number of atoms and same width of the actual distribution. Obviously, the actual distribution cannot fit well to a Gaussian. This also means that the integrable system TG gas cannot be thermalized, although it can evolve into a steady-state.

What is the steady-state? How can it be described? The steady-state of the integrable system can be described by a new ensemble, which is named the fully constrained thermodynamic ensemble [21]. It is defined by all of its integrals of motion for the former 1D hard-core boson (TG gas) system,

$$\hat{\rho}_{FC} = Z_{FC}^{-1} exp \left[-\sum_k \lambda_k \hat{f}^F(k) \right], \qquad (14.15)$$

where $Z = \prod_k (1 + e^{-\lambda_k})$ is a normalization factor and the integrals of motion $\hat{f}^F(k)$ have been previously defined. In the following, the numerical results show that this ensemble is the correct description of the steady-state of the integrable system, rather than the thermal state. This can be verified by the momentum distribution of the steady-state in Figs. 14.10 and 14.11 [21].

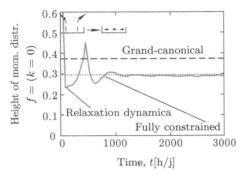

Figure 14.10.   The momentum distribution of the relaxation dynamics. From Ref. [21].

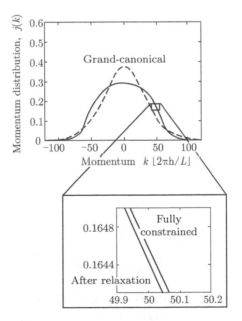

Figure 14.11.   The momentum distribution of the steady-state. From Ref. [21].

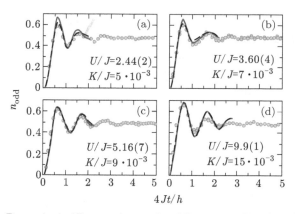

Figure 14.12.   Dynamics in 1D strongly correlated Bose gases. From Ref. [31]. The solid lines are simulated by the t-DMRG algorithm without free parameters and the dashed line denotes the same simulation with the next-nearest hopping ($J_{NNN}/J \simeq 0.12$ in (a), 0.08 in (b), 0.05 in (c) and 0.03 in (d)). It is worth noting that this controlled coherent evolution is beyond the calculation ability of current classical computers. If this can be used as a toolbox for calculation, its ability has exceeded that of classical machines.

Although the integrable system cannot be thermalized, that does not mean that the non-integrable system can be thermalized. Some numerical results show that there are also some non-integrable systems that cannot be thermalized [22]. The recent experiment with the strongly correlated 1D Bose gas, which is non-integrable, indicates that there is indeed a closed system that can be thermalized [31]. The experiment is based on the 1D BH model

$$H = \sum_j \left[-J(\hat{a}_j^\dagger \hat{a}_j + H.C.) + \frac{U}{2}\hat{n}_j(\hat{n}_j - 1) + \frac{K}{2}\hat{n}_j j^2 \right], \qquad (14.16)$$

in which we use J as the hopping rate to distinguish time t. $K = m\omega^2 d^2$ where m is the particle mass, d is the lattice spacing and ω is the trapping frequency of the external harmonic trap; the experiment includes three steps:

- When $t = 0$, initializing the system to the state $|\cdots, 1, 0, 1, 0, 1, \cdots\rangle$ with even sites occupied and no hopping present in the chain.
- Detuning the parameters J, U and K to some positive values and the system will begin the non-equilibrium dynamics with the Hamiltonian H
- Supressing hopping and reading out the state $|\psi(t)\rangle$.

The experimental results agree very well with the numerical simulation and show that the system can be thermalized.

It is very important to reconsider our statistic mechanics. It is supposed that any closed system with sufficient degrees of freedom can be thermalized. This assumption has its own limit. The thermalized state is supposed to be a mixed state (described by a density matrix), while the unitary evolution of a closed system from an initial pure state should be another pure state. If the thermalization process really exists, then the decoherence process should be introduced. However, the origin of the decoherence process must be determined. Much attention has been focused on this problem and some new theories have been proposed, such as the eigenstate thermalization hypothesis [32, 33], to attack it. However, it remains an open question.

References

[1] M. Greiner, M. O. Mandel, T. Esslinger, T. Hnsch and I. Bloch. *Nature* **415**, 39 (2002).
[2] S. R. White and A. E. Feiguin. *Phys. Rev. Lett.* **93**, 076401 (2004).
[3] A. J. Daley, C. Kollath, U. Schollwöck and G. Vidal. *J. Stat. Mech. Theor. Exp.* P04005 (2004).
[4] D. Gobert, C. Kollath, U. Schollwöck and G. Schutz. *Phys. Rev. E* **71**, 036102 (2005).
[5] M. Zwolak and G. Vidal. *Phys. Rev. Lett.* **93**, 207205 (2004).
[6] G. Vidal. *Phys. Rev. Lett.* **98**, 070201 (2007).
[7] J. Jordan, R. Orus, G. Vidal, F. Verstraete and J. 1. Cirac. *Phys. Rev. Lett.* **101**, 250602 (2008).
[8] F. Verstraete and J. I. Cirac. arXiv:cond-mat/0407066.
[9] I. Affleck, T. Kennedy, E. H. Lieb and H. Tasaki. *Comm. Math. Phys.* **115**, 477–528 (1988).
[10] M. Fannes, B. Nachtergaele and R. F. Werner. *Comm. Math. Phys.* **144**, 3 (1992).
[11] S. Ostlund and S. Rommer. *Phys. Rev. Lett.* **75**, 19 (1995).
[12] M. Suzuki. *Phys. Lett. A* **146**, 6 (1990); *J. Math Phys.* **32**, 2 (1991).
[13] A. T. Sornborger and E. D. Stewart. arXiv:quant-ph/9809009.
[14] R. A. Horn and C. R. Johnson. *Matrix Analysis.* Cambridge University Press, Cambridge (1990).
[15] G. Vidal, J. I. Latorre, E. Rico and A. Kitaev. *Phys. Rev. Lett.* **90**, 227902 (2003).
[16] C. Kollath, A. Laeuchli and E. Altman. *Phys. Rev. Lett.* **98**, 180601 (2007).
[17] M. Greiner, M. O. Mandel, T. Hansch and I. Bloch. *Nature* **419**, 51 (2002).
[18] K. Huang. *Statistical Mechanics.* John Wiley & Sons, New York (1990).

[19] N. Linden, S. Popescu, A. J. Short and A. Winter. *Phys. Rev. E* **79**, 061103 (2009).

[20] O. Lychkovskiy. *Phys. Rev. E* **82**, 011123 (2010).

[21] M. Rigol, V. Dunjko, V. Yurovsky and M. Olshanii. *Phys. Rev. Lett.* **98**, 050405 (2007).

[22] C. Gogolin, M. P. Muller and J. Eisert. *Phys. Rev. Lett.* **106**, 040401 (2011).

[23] B. Sutherland. *Beautiful Models*. World Scientific, Singapore (2004).

[24] V. E. Korepin and F. H. L. Essler. *Exactly Solvable Models of Strongly Correlated Electrons*. World Scientific, Singapore (1994).

[25] M. Giardeau, *J. Math. Phys.* **1**, 516 (1960).

[26] E. H. Lieb and W. Liniger. *Phys. Rev.* **130**, 1605 (1963).

[27] B. Paredes, A. Widera, V. Murg, O. Mandel, S. Folling, J. I. Cirac, G. V. Shlyapnikov, T. W. Hansch and I. Bloch. *Nature* **429**, 277 (2004).

[28] T. Kinoshita, T. Wenger and D. S. Weiss. *Science* **305**, 1125 (2004).

[29] I. Bloch, J. Dalibard and W. Zwerger. *Rev. Mod. Phys.* **80**, 885 (2008).

[30] T. Kinoshita, T. Wenger and D. S. Weiss. *Nature* **440**, 900 (2006).

[31] S. Trotzky, Y-A. Chen, A. Flesch, I. P. McCulloch, U. Schollwock, J. Eisert and I. Bloch. *Nat. Phys.* **8**, 325 (2012).

[32] M. Rigol, V. Dunjko and M. Olshanii. *Nature* **452**, 854 (2008).

[33] M. Srednicki. arXiv:cond-mat/9410046; M. Srednicki, *J. Phys. A* **29**, 75 (1996).

15
Disordered Systems

As an ubiquitous factor in solid-state systems due to random defects and dislocations, disorder introduces various exotic many-body phenomena that have attracted great attention in condensed matter physics. The most famous example is the Anderson localization in a free electron system [1]. In non-interacting systems, it is widely believed that localization occurs in one and two dimensions for an arbitrarily small disorder, whereas there is a critical value for the strength of disorder in a three-dimensional system. However, there is little knowledge about the interplay between disorder and interaction, in particular strong interaction, in many-body systems. In addition, there are several difficulties hampering experimental investigation on this subject using conventional condensed matter systems [2]. First, disorders cannot be easily controlled or experimentally fine-tuned. Second, samples prepared using the same method acquire some distribution of disorders. To characterize their properties, we need to perform measurements for as many samples as possible and average the results. This requires a large number of samples, which is quite demanding in time and labor, and usually very expensive. Third, in some disordered systems, there may be a large number of low-energy excitations, which makes a quantitative investigation more subtle and difficult. Besides, the most important problem is to investigate the interplay between disorders and interactions in a many-body system. This means that we need to control the disorder and the interaction together, which is almost impossible in conventional materials.

Many of these difficulties can be overcome in the optical lattice platform. Specifically, the random potentials that induce disorders can be well controlled, such that the strength of disorder can be experimentally tuned.

Besides, every realization of disorder can be easily obtained and the interactions between atoms can also be tuned by means of Feshbach resonances.
Thus, we can investigate the interplay of interaction and disorder in a same
system. In addition, the systems of atomic gases in optical lattices can be
used to investigate novel problems that were not accessible before. As an
example, the quantum statistics can be chosen by using ultracold bosons or
fermions, and we can research the relations between disorder and quantum
statistics. In addition, the controllability of the trapping potentials, and
therefore the effective dimensionality of the system, makes it possible to
investigate the role of dimensionality in disordered systems.

The atomic disordered systems in optical lattices have their own intrin-
sic disadvantage: the atomic systems are small compared to the conventional
condensed matter system. Despite this disadvantage, the optical lattice is
an ideal platform for studying disordered systems.

15.1.   Disorder in free space

There are several ways to generate random potentials in optical lattices
[3–9]. In general, the distribution of a random potential can be of the form
of any function, such as a Gaussian. Here, we consider only the uniform
distribution case. The simplest way to obtain such a distribution is from a
speckle pattern [3–6]. The speckle field is a stationary and coherent light
field with highly disordered intensity and phase distributions. A schematic
illustration of the setup used to achieve the optical speckle potential is
shown in Fig. 15.1 [2]. As an alternative, disorders can also be realized by
implementing within a 1D quasi-periodic, incommensurate lattice [7–9].

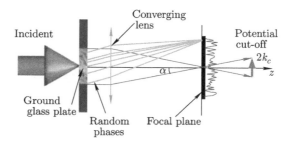

Figure 15.1.   An illustration of the experimental setup to produce disorders with a
speckle field. From Ref. [2].

One of the most well-known phenomenon induced by disorders is the Anderson localization. In his seminal paper in 1958 [1], Anderson predicted that a localized state can be stabilized in free electronic systems when subjected to strong enough disorders, such that the system undergoes a metal-insulator transition to become a so-called Anderson insulator. To observe such an exotic phase, the system should satisfy several conditions:

- The Anderson localization occurs in free systems, thus the internal interactions and all perturbations, such as time-dependent fluctuations of the medium, must be eliminated.
- A weak-enough disorder must be used so that interference effects (the origin of the Anderson localization) can dominate over classical trapping in the potential minima.
- Supression of transport, which can also arise from classical trapping, and exponential localization must be demonstrated.

In comparison to conventional condensed matter systems, the ultracold atomic gases have many advantages. First, the system are very dilute, thus the interaction between the atoms can be negligible. Second, the atoms are well isolated from the environment and the fluctuation can be substantially suppressed. Third, the disorder can be controlled by the intensity of the speckle field to be sufficiently weak. Finally, the localization can be found by the direct imaging of atomic density profiles. Therefore, the ultracold atom system is a good platform to directly observe the Anderson localization. There are two different schemes to observe this effect: a transport scheme [10] and a static scheme [11].

- *Transport scheme:* In the transport scheme, a very dilute weakly interacting BEC is loaded in a trap. The trap is suddenly switched off at some time, denoted as $t = 0$. Then, the condensate will expand through a given guide. The disordered potential that is created by the optical speckle will be present in this guide. During the expansion of the condensate, the number density decreases fast, such that the already weak interaction effect becomes negligible soon after the release. As a result, a free gas is produced in the disorder potential with all conditions fulfilled to realize the Anderson localization. The transport characteristics can be extracted from the expansion dynamics. In Fig. 15.2, we show the experimental results obtained by Billy *et al.* [10], where theoretical predictions match the data points in a quantitative level.

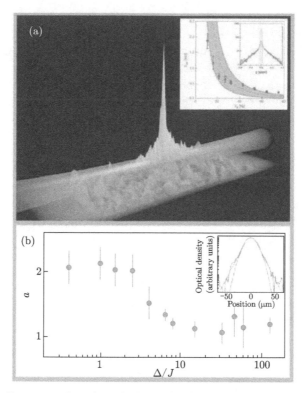

Figure 15.2.   Demonstration of an Anderson localization in an atomic system. From Ref. [10].

- *Static scheme:* In contrast, the static scheme does not require an expansion of the BEC. Instead, the interaction between the trapped atoms is switched off by tuning through a Feshbach resonance [11]. The disorder in this scheme can be formed by a 1D quasi-periodic, incommensurate lattice.

15.2.   Disorder in an optical lattice

The Anderson localization of non-interacting systems is well understood. However, the disorder in interacting systems, especially in strongly interacting systems [2], is still a significant challenge. The interplay between the disorder and the interaction is still not clear. The Bose–Hubbard (BH) model is a simple interaction model and it has already been realized in the optical lattice. It can be used as a tool to investigate this challenging problem. In addition, the disordered BH (DBH) model has undergone significant theoretical study. The disorder in this model induces some interesting new phases [12]. The Hamiltonian of the BH model with disorder can be written as

$$H = -\sum_{\langle ij \rangle}(t_{ij}b_i^\dagger b_j + \text{H.C.}) + \frac{U}{2}\sum_i n_i(n_i - 1) + \sum_i \mu_i n_i, \qquad (15.1)$$

where b_i and $b_j^\dagger$ are bosonic field operators on site i, H.C. stands for Hermitian conjugate, and

$$\mu_i \equiv \int d\mathbf{r} w^*(\mathbf{r} - \mathbf{R}_i)[V_r(\mathbf{r}) + V_T(\mathbf{r})]w(\mathbf{r} - \mathbf{R}_i) = \varepsilon_i - \widetilde{\mu}_i \qquad (15.2)$$

denotes the local chemical potential with $V_r(\mathbf{r})$ being the random potential with some distribution, $V_T(\mathbf{r})$ being the lattice potential and $w(\mathbf{r} - \mathbf{R}_i)$, the Wannier function of site R_i as in the conventional BH model. Here, we define a site-dependent chemical potential $\widetilde{\mu}_i$, which by assumption acquiring the same distribution at every site. The simplest distribution is the uniform distribution between $-\Delta$ and Δ and this distribution of μ_i can be described by a single parameter Δ that characterizes the strength of disorder.

Theoretical investigations suggest four possible phases for such a disordered BH model, including superfluid (SF), Mott insulator (MI), Bose glass (BG) and Anderson glass [13]. The SF and MI phases are well understood in the BH model. The two glass phases are new phenomena in disordered systems that resulted from the interplay between disorder and interaction. The two glass phases are distinguished in the following aspects. First, the BG phase appears at the strong coupling region where disorder and interactions tend to cooperate. On the contrary, the Anderson glass phase is supposed to appear in the weak coupling region where the interactions compete with the disorder and delocalize the bosons. Second, the distributions of the bosonic number density within these two phases are substantially different. In the BG phase, the density is reasonably uniform, while the boson density correlations are expected to exponentially decay in the Anderson glass phase.

The BG phase has been confirmed by many analytical and numerical calculations and been witnessed experimentally [15–17]. However, the Anderson glass is still not confirmed, although this is claimed in some calculations [13].

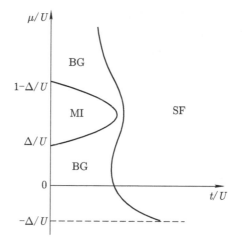

Figure 15.3. The phases of a disordered Bose–Hubbard (BH) model predicted by the mean-field theory.

A general phase diagram of a one-dimensional BH model with disorder is first given by Fisher *et al.* [12] by the mean-field theory, as shown in Fig. 15.3. In this diagram, the BG phase is completely separated from the adjacent SF and MI phases. However, this mean-field phase diagram is too rough; it cannot answer whether the MI phase can move directly to the SF phase without going through the BG phase. To clarify the details, more precise numerical calculations are needed.

For a 3D situation, there are numerical calculations indicating the absence of a direct transition from the SF to the MI phase [18], as illustrated in Fig. 15.4. This conclusion is based on the following two theorems [12, 19]

Theorem 15.1. *If the bound Δ on the disorder strength is larger than the half-width of the energy gap $E_g/2$ in the ideal MI phase, the system is inevitably compressible; i.e., the transition is to the BG insulator and not the MI whenever*

$$\Delta > E_g/2. \tag{15.3}$$

Theorem 15.2. *There is a non-zero compressibility on the superfluid-insulator critical line and in its neighborhood, for models that have disordered on-site potentials.*

We define the parameter Δ_c as the critical value of Δ from SF to MI. From the former two theorems, we can conclude that if there is no

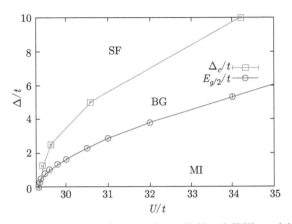

Figure 15.4.   The phase diagram of the 3D Bose–Hubbard (BH) model with disorder. From Ref. [18].

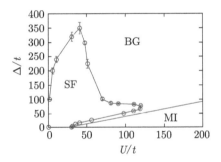

Figure 15.5.   Another phase diagram of the 3D Bose–Hubbard (BH) model with disorder. From Ref. [20].

direct transition, the parameter satisfies $\Delta_c > E_g/2$. The Quantum Monte Carlo (QMC) calculation shows that the parameters do have the relation $\Delta_c > E_g/2$. Another calculation of the 3D disordered BH model [20], which directly calculates the phase boundary of the different phases, suggests a phase diagram at the unity filling, as shown in Fig. 15.5.

This diagram is consistent with the former claim. All these calculations show that the SF state cannot undergo a direct transition to the MI state. However, other calculations, such as the stochastic mean field and replica theory, indicate a possibility of a direct transition from the MI to SF. Thus, results from the low dimensional (1D and 2D) disordered BH model are more controversial and remain an open question. Because of the successful

realization of the phase transition from SF to MI in an optical lattice, we
hope to obtain new insights or even completely solve this problem within
this novel platform.

If we were to control the disorder, we could completely simulate and
manipulate the interaction and disorder. As mentioned before, the disorder
can be generated by the speckle field. The speckle field will modify the
entire potential of the optical lattice, as shown schematically in Fig. 15.6
[16]. An example of the total potential in the experiment is illustrated in
Fig. 15.7.

We have previously mentioned that the parameters of the BH model in
the optical lattice depend on the intensity of the field. Now, the intensity
of the field is modified by the speckle field and is randomly distributed
in experiments. The parameters of the BH model (not just the chemical potential) must be modified and have some distributions. The typical
distributions in the experiment are shown in Fig. 15.8. Notice that these
distributions are determined for $s = 14E_r$ (to satisfy the deep potential

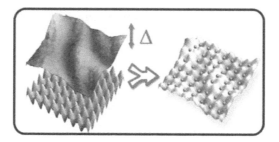

Figure 15.6.   An image of the speckle field modifying the optical lattice. From Ref. [16].

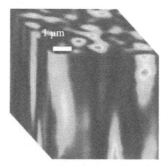

Figure 15.7.   An experimental example of the modified potential. From Ref. [16].

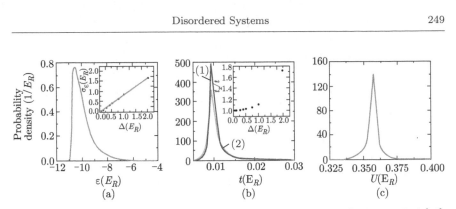

Figure 15.8.   The distribution of different parameters with a speckle field. From Ref. [16].

condition) and $\Delta = 1E_r$ (the disorder strength). Figure 15.8(a) shows the real distribution of the on-site energies ε (corresponding to the disorder term in the DBH model). The inset in this figure shows the dependence of the on-site energy's standard deviation on the disorder strength Δ, which can be almost linearly fitted. Figure 15.8(b) shows the distribution of the hopping parameter t in different directions (1) for the x direction and (2) for the z direction). The inset of this figure shows the ratio of the hopping rate between the different directions (x and z) over the disorder strength Δ. Figure 15.8(c) shows the distribution of the interaction parameter U.

All these figures show that the speckle field broadens all the distributions. The on-site parameter ε and hopping parameter t are more sensitive to the speckle field than the interaction parameter U, and the relative scale is determined by the width of the distribution over their mean. We recall that the hopping rate t is more sensitive to the intensity than the interaction U in the standard BH model. This distribution indicates that the system cannot be exactly described by the former DBH model: there is still some disorder in the hopping parameter, though the disorder in the interaction can be neglected. We do not expect to determine all the characteristics of the DBH model with this modified model. However, we assume that these two models have similar physical properties, which can give important insights into the DBH model.

In the experiment, the effect of the disorder on the fraction of the condensate [16] (which is closely related to but not the same as the fraction of the SF) is measured (see Fig. 15.9). The fraction can give indirect information about the transformation between the MI and SF with disorder. The fraction of the condensate can be found in the TOF image. From this image, we can define N_0 as the condensate part, which is the narrow central

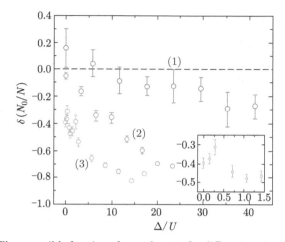

Figure 15.9.   The reversible fraction of a condensate for different regions. From Ref. [16]. The parameters of the different data are $s = 6$ (SF with negligible depletion, see (1)), $s = 12$ (SF with strong depletion, see (2)) and $s = 14$ (MI and SF with depletion, see (3)). The data shows the relation of the reversible fraction of the condensate for different regimes with the disorder Δ/U. The dashed line shows the case where the disorder does not change the condensate. The inset is a close-up of curve (3) with low Δ.

component of the profiles. In Fig. 15.9, the different s values correspond to the different regimes for the pure BH model with a parabolic confining potential.

Generally, the disorder in the site and hopping causes the depletion of the condensate in the DBH model with a harmonic trap. This result is indirect evidence that the disorder will cause the superfluid region to change to another phase. The present experiment cannot give us a clear answer to the puzzles of disordered systems. However, it is a promising and novel way forward for these puzzles.

References

[1] P. W. Anderson. *Phys. Rev.* **109**, 1492 (1958).
[2] L. Sanchez-Palencia and M. Lewenstein. *Nat. Phys.* **6**, 87 (2010).
[3] J. E. Lye, L. Fallani, M. Modugno, D. Weirsma, C. Fort and M. Inguscio. *Phys. Rev. Lett.* **95**, 070401 (2005).
[4] C. Fort, L. Fallani, V. Guarrera, J. Lye, M. Modugno, D. S. Wiersma and M. Inguscio. *Phys. Rev. Lett.* **95**, 170410 (2005).
[5] T. Schulte, S. Drenkelforth, J. Kruse, W. Ertmer, J. Arlt, K. Sacha, J. Zakrzewski and M. Lewenstein. *Phys. Rev. Lett.* **95**, 170411 (2005).

[6] D. Clement, A. F. Varon, M. Hugbard, J. A. Retter, P. Bouyer, L. Sanchez-Palencia, D. M. Gangardt, G. V. Shlyapnikov and A. Aspect. *Phys. Rev. Lett.* **95**, 170409 (2005).

[7] B. Damski, J. Zakrzewski, L. Santos, P. Zoller and M. Lewenstein. *Phys. Rev. Lett.* **91**, 080403 (2003).

[8] R. Roth and K. Burnett, *J. Opt. B: Quantum Semiclass. Opt.* **5**, S50 (2003).

[9] R. Roth and K. Burnett. *Phys. Rev. A* **68**, 023604 (2003).

[10] J. Billy, V. Josse, Z. Zuo, A. Bernard, B. Hambrecht, P. Lugan, D. Clément, L. Sanchez-Palencia, P. Bouyer and A. Aspect. *Nature* **453**, 891 (2008).

[11] G. Roati, C. D'Errico, L. Fallani, M. Fattori, C. Fort, M. Zaccanti, G. Modugno, M. Modugno and M. Inguscio. *Nature* **453**, 895 (2008).

[12] M. P. A. Fisher, P. B. Weichman, G. Grinstein and D. S. Fisher. *Phys. Rev. B* **40**, 546(1989).

[13] R. T. Scalettar, G. G. Batrouni and G. T. Zimanyi. *Phys. Rev. Lett.* **66**, 3144 (1991).

[14] M. Lewenstein, A. Sanpera, V. Ahufinger, B. Damski, A. Sen(De) and U. Sen. *Adv. Phys.*, **56**, 243 (2007).

[15] M. Kohl, H. Moritz, T. Stoferle, K. Gunter and T. Esslinger. *Phys. Rev. Lett.* **94**, 080403 (2005).

[16] M. White, M. Pasienski, D. McKay, S. Zhou, D. Ceperley and B. De Marco. *Phys. Rev. Lett.* **102**, 055301 (2009).

[17] M. Pasienski, D. McKay, M. White and B. De Marco. *Nat. Phys.* **6**, 677 (2010).

[18] L. Pollet, N. V. Prokofev, B. V. Svistunov and M. Troyer. *Phys. Rev. Lett.* **103**, 140402 (2009).

[19] J. K. Freericks and H. Monien. *Phys. Rev. B* **53**, 2691(1996).

[20] V. Gurarie, L. Pollet, N. V. Prokof'ev, B. V. Svistunov and M. Troyer. *Phys. Rev. B* **80**, 214519 (2009).

16
Simulation of Spin Systems

16.1. General phases of spin systems

The study of spin systems remains an important field of condensed matter physics that is closely related to magnetism phenomena, ever since the beginning of the quantum theory. There are many famous spin models that play essential roles in the understanding of many-body physics. In addition, some of the spin models, such as the 1D quantum Ising model [1] and the 2D classical Ising model [2], can be solved exactly. These solvable models are often used as standard models to test new methods, either analytic or numerical, before they can be safely used to tackle unresolved problems.

Although there are many different spin models, the forms of interactions between spins are limited. Actually, there are just two typical interactions. The simpler one is the Ising interaction

$$H_I = J \sum_{\langle ij \rangle} \sigma_i^\mu \sigma_j^\mu, \qquad (16.1)$$

where $\sigma^{x,y,z}$ are Pauli matrices and μ represents one spatial coordinate chosen from x, y or z. For the case of $J > 0$, this model is called an antiferromagnetic model, while in the case of $J < 0$, it is called the ferromagnetic (F) model. The other interaction is the Heisenberg interaction [3]

$$H_H = J \sum_{\langle ij \rangle} \vec{\sigma}_i \cdot \vec{\sigma}_j. \qquad (16.2)$$

Generally, i and j are the nearest neighbors in the lattice. In some special cases, they may be next nearest neighbors.

Despite the simple form of interactions, spin systems can have rich phase diagrams [4–6]. In particular, there are some novel exotic phases, such as a spin liquid, which may be closely related to high T_c superconductivity, and these phases are of great interest in condensed matter physics. One can encounter these typical phases in spin systems:

• Neel order

This is a standard up-down-up-down state. If the lattice sites can be divided into two sublattices, A and B, and every site in sublattice A is only surrounded by the sites in sublattice B and every site in sublattice B is only surrounded by the sites in sublattice A, then the lattice is called a bipartite lattice. The Neel state is defined when on the bipartite lattice, the spins in sublattice A are up and the spins in sublattice B are down. An illustration of this state on a square lattice is shown in Fig. 16.1. Notice this state breaks the rotational and translational symmetry absolutely. It has a long-range order and leads to gapless spin wave excitations.

• Valence bond solids (VBS)

The simplest model that has the VBS ground state is the one-dimensional Majumdar–Ghosh model [7]. There is an even number of sites in this model. With periodic boundary conditions, the Hamiltonian can be written as

$$H = \sum_i \frac{2}{3}(\mathbf{S}_i \cdot \mathbf{S}_{i+1} + \mathbf{S}_{i+1} \cdot \mathbf{S}_{i+2} + \mathbf{S}_i \cdot \mathbf{S}_{i+2}). \qquad (16.3)$$

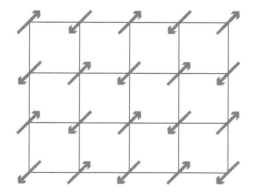

Figure 16.1.   The configuration of the Neel state.

The ground state of this model is

$$|\psi_{\mathrm{MG}}^{G}\rangle = \prod_{i=1}^{N/2}(|0\rangle_{2i}|1\rangle_{2i+1} - |1\rangle_{2i}|0\rangle_{2i+1})/\sqrt{2}. \qquad (16.4)$$

If we denote the singlet state of a pair of spins $(|0\rangle|1\rangle - |1\rangle|0\rangle)/\sqrt{2}$, which is also called a dimer, the ground state can be represented as in Fig. 16.2, with dimers indicated by solid lines.

Another famous one-dimensional model with the VBS ground state is the Affleck–Kennedy–Lieb–Tasaki (AKLT) model [8–10]

$$H = \sum_{i} \frac{1}{2}\mathbf{S}_i \cdot \mathbf{S}_{i+1} + \frac{1}{6}(\mathbf{S}_i \cdot \mathbf{S}_{i+1})^2 + \frac{1}{3}, \qquad (16.5)$$

where $\mathbf{S}$ is the spin operator with spin 1. This model plays important roles in many-body physics, especially in the development of the DMRG method [8–10] and the Tensor Network algorithm. The state of this model has a very famous topological edge state, as shown in Fig. 16.3.

A typical configuration of dimers in a VBS in a two-dimensional lattice is shown in Fig. 16.4, where the characteristics of the VBS state can be

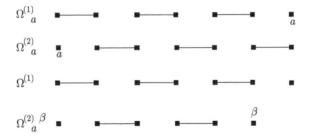

Figure 16.2.   The configuration of the ground state of the Majumdar–Ghosh model.

Figure 16.3.   The configuration of the ground state of the AKLT model. As per the former definition, the line denotes a singlet, while the circle corresponds to a projector because the product space of two spins is four dimensions and the dimensionality of the Hilbert space of spin 1 is only three.

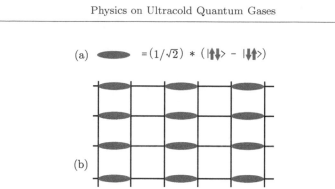

Figure 16.4.   The configuration of the VBS state.

easily extracted. Similar to a solid, the array of singlets in a VBS is ordered and covers all sites in the lattice. Therefore, if we consider the dimer as a whole, there is a long-range order in dimer correlations. The dimer lattice has translational symmetry [11]. The system is gapped and correlation functions (general two operator correlations, not dimer correlations) decay exponentially.

• **Spin liquid**

This exotic state is extremely interesting in spin systems because it is possibly related to high T_c superconductivity. It seems that the spin liquid will exceed the Landau–Wilson paradigm, which is the standard theory of the continuous phase transition theory [12]. The spin liquid has no apparent break in symmetry, which cannot be described by the Landau–Wilson theory. In particular, there is a topological spin liquid, for which the degeneracy of the ground state depends on the topology of the underlying lattice. The important example of this topological spin liquid is the Kitaev model [13], which we will describe in detail later. When a VBS, which is a dimer solid, starts to melt due to the quantum fluctuations introduced into the system, there will be some unpaired spins that will move to any of the sites. Therefore, the resulting state will be the superposition of the random paired dimer states, which is the Resonating Valence Bond (RVB) liquid [14]. The melting process is illustrated in Fig. 16.5. The RVB liquid is gapless and has many low-energy excitations. An example of the RVB liquid is the Heisenberg antiferromagnet in the trimerized Kagome lattice [15–27], which will also be introduced later.

Due to the exotic nature and importance of the spin liquid state, significant work is needed to investigate the properties of this phase. The

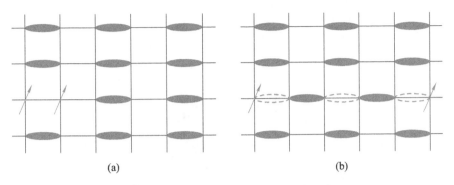

Figure 16.5.   The melting process of VBS.

frustrated spin systems are the most promising systems that may have a
spin liquid ground state. The frustrated spin systems [5, 6] also deserve
intensive study on their own merits. Generally, the spin system can be
well understood with numerical calculations by the Monte Carlo methods.
However, the frustrated spin systems cannot be calculated by these methods
due to the sign problem. In particular, 2D frustrated systems are difficult
to understand. The possible methods to understand the frustrated systems
include DMRG, PEPS and some mean-field theories. The DMRG works
well for 1D (or quasi-1D) systems, but not for higher dimensions, with
the exception of some limited boundary conditions, such as the cylinder
condition to form a quasi-1D system [26]. Therefore, this is also a possible
means to study quasi-1D systems and obtain more insights into the 2D
system. The PEPS method is a natural extension of the DMRG algorithm
to the 2D case.

The simplest frustrated system is the Ising model on a triangle, as
shown in Fig. 16.6. This mode is not frustrated on the regular square lattice. However, when it is on a triangle, it is frustrated. If one spin is in the up
state ($|1\rangle$), one of its neighbors can be determined to be in the down state
($|0\rangle$) to minimize the energy of the Ising interaction. However, the state
of third spin cannot be $|0\rangle$ or $|1\rangle$ because neither state can minimize the
energy of the entire Ising interaction. Thus, the third spin will be frustrated
to determine its state. In other words, whether the third spin is $|1\rangle$ or $|0\rangle$,
the energy of the system will be the same. There are many spin configurations that have the same energy, which means they are degenerate. It is
deserving of mention that the direction of spins discussed here are classical
concepts because the quantum fluctuation of the spins are not considered.

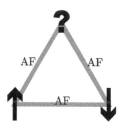

Figure 16.6.   The simplest frustrated system on a triangle lattice.

Figure 16.7.   The Kagome lattice.

Generally, there are two elements in the spin systems: lattices and inter-action. Therefore, the frustration in the spin systems can also be divided into two different types according to the former elements. The first type of frustration comes from the geometry of lattices with the regular nearest interaction. The typical "frustrated" lattices are the Kagome lattice and the triangle lattice [4].

The Kagome lattice is as shown in Fig. 16.7. The Heisenberg Hamilto-nian on this lattice is

$$H_{kag} = J \sum_{\langle ij \rangle} \vec{\sigma}_i \cdot \vec{\sigma}_j + J' \sum_{\langle ij \rangle} \vec{\sigma}_i \cdot \vec{\sigma}_j, \qquad (16.6)$$

where the positive parameters J and J' are the interaction strength of the spins belonging to different triangles, as defined in Fig. 16.8. There is evidence to show that the ground state of this system is in a spin liquid state. First, in the fully trimerized limit $J'/J = 0$, the ground state is in

Figure 16.8. The definition of the parameters J and J'. The interaction coupling in the parent triangle that contains three small triangles at the corner is J and the coupling in the other inverted parent triangle is J'.

the short-ranged RVB subspace and there is one singlet in each triangle. Second, if the parameters in J'/J are very small, then the mean-field theory shows that the spectrum, the form and the number of singlets consist of the short-ranged RVB ground states [15–25]. For the general case, numerical simulations show that the energy gap between the ground state and the lowest triplet state, if any, is also very small. Furthermore, the low-lying singlets, whose numbers scale with 1.15^N, where N is the number of spins, fill the gap. All these traits suggest that the ground state of this model may be described by the RVB states. Very recently, the spin liquid state has been strongly supported by experiments [27].

The other type of frustration comes from the next nearest interaction with the regular lattice, such as a square lattice. The simplest one is the J_1–J_2 model on the square lattice with the Hamiltonian

$$H_{J_1 J_2} = 2J_1 \sum_{\langle ij \rangle} \vec{\sigma}_i \cdot \vec{\sigma}_j + 2J_2 \sum_{\langle\langle ij \rangle\rangle} \vec{\sigma}_i \cdot \vec{\sigma}_j, \qquad (16.7)$$

where $\langle ij \rangle$ and $\langle\langle ij \rangle\rangle$ represent the nearest and the next nearest neighbors, respectively. With different parameters of J_1 and J_2, the model will have different phases. In the limiting case of $J_2 \ll J_1$, the system is in the Neel state. On the opposite limit of $J_2 \gg J_1$, the system is in the collinear state, which is characterized as

$$|\psi_{J_2 \gg J_1}^G\rangle = \prod_{i \text{ odd} j} \prod |1\rangle_{ij} \prod_{\text{even} j} |0\rangle_{ij}, \qquad (16.8)$$

with i and j indices of rows and columns, respectively.

However, in the strongly frustrated region, where $0.4 \lesssim J_2/J_1 \lesssim 0.6$, the situation is far from conclusive. Different calculations give contradictory conclusions [28–35]. The semiclassical limit shows a direct transition from the Neel phase to the collinear phase. Meanwhile, many calculations suggest that the new phase may have the VBS dimer configuration with a long-range order. Recently, there are calculations to support the existence of a spin liquid phase. In particular, the exact diagonalization (ED) calculation [28–34] shows that the new phase is a spin liquid VBS. The spin liquid region is $0.38 \lesssim J_2/J_1 \lesssim 0.6$. The variational approach suggests that the ground state of the system is a spin liquid RVB [28–34]. Besides, a new algorithm based on the PEPS state [35] and the results from the DMRG algorithm [36] also support the existence of the spin liquid phase between the Neel phase and the bilinear phase. Thus far, the ground state of the highly frustrated region is not yet completely understood. However, this system is a good candidate to investigate the spin liquid phase.

Unlike the J_1–J_2 frustrated model, the spin liquid phases can only be supported by some numerical methods. At this moment, strong evidence confirm the existence of a spin liquid state in the Kitaev model. The Kitaev model is defined on a honeycomb lattice and the Hamiltonian is defined as [13]

$$H_{\text{Kitaev}} = -J_x \sum_{x\text{link}}' \sigma_j^x \sigma_k^x - J_y \sum_{y\text{link}} \sigma_j^y \sigma_k^y - J_z \sum_{z\text{link}} \sigma_j^z \sigma_k^z, \qquad (16.9)$$

where the interactions between different directions are different and they can be expressed in Fig. 16.9.

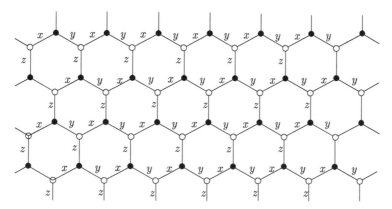

Figure 16.9.   The Kitaev model on a honeycomb lattice. Edges with different directions correspond to different interactions.

If we introduce the Majorana operators into this model, the spin operators can be represented as

$$\sigma^x = ib^x c, \sigma^y = ib^y c, \sigma^z = ib^z c, \tag{16.10}$$

where b^x, b^y, b^z and c are Majorana operators. The Majorana operators c_i satisfy the relations

$$c_k^\dagger = c_k, \{c_k, c_l\} = 2\delta_{kl}. \tag{16.11}$$

Then the Kitaev model can be represented by the Majorana form as in Fig. 16.10. The Hamiltonian can be represented by the new operators as

$$\hat{H}_{\text{Kitaev}} = \frac{i}{4} \sum_{\langle j,k \rangle} \hat{A}_{jk} c_j c_k \tag{16.12}$$

and

$$\hat{A}_{jk} = 2J_{d(j,k)} \hat{\mu}_{jk} \hat{\mu}_{jk} = ib_j^{d(j,k)} b_k^{d(j,k)}, \tag{16.13}$$

where $d(j,k) = x, y, z$ is the direction of the link j and k.

The Fock space, supported by the Majorana fermions in the Hamiltonian $\hat{H}_{\text{Kitaev}}$, is larger than the real physical space. There are some nonphysical freedoms in the Majorana fermion space. The physical space can

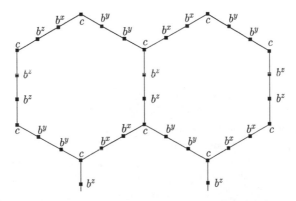

Figure 16.10.   Representation of the Kitaev model with Majorana operators.

be reached by some limitation on the Majorana operators, such as

$$D_j|\psi\rangle = |\psi\rangle, \qquad (16.14)$$

where $D_j = b_j^x b_j^y b_j^z c$. This condition can also be realized by a projector, which is more convenient for practical use. The projector is defined as

$$P - \prod_j \left(\frac{1 + D_j}{2}\right). \qquad (16.15)$$

In other words, the physical space of the Kitaev model can be found from the Majorana Fock space by projecting (defined by the projector) into the physical space. If we can exactly solve the Majorana Kitaev model, then we can obtain the full physics of the Kitaev model.

Given the characteristics of the Majorana operators, the operator $\hat{A}_{jk}$ commutes with itself. Therefore, $\hat{A}_{jk}$ commutes with the whole Majorana Hamiltonian $\hat{H}_{Kitaev}$. Thus, they are the conversion quantities of the Majorana fermion system and we can fix $\mu_{jk} = \pm 1$. These conversion parameters can be used to define an orthogonal decomposition of the full space

$$\zeta = \bigoplus_\mu \zeta_\mu, \qquad (16.16)$$

where $|\Psi\rangle \in \zeta_\mu$ iff $\hat{\mu}_{jk}|\Psi\rangle$ for any j and k. In a given subspace ζ_μ, the corresponding Majorana type Kitaev model is a quadratic Hamiltonian

$$\hat{H}_\mu = \frac{i}{4} \sum_{\langle j,k \rangle} A_{jk} c_j c_k, \qquad (16.17)$$

where $A_{jk} = 2J_d(j,k)\mu_{jk}$ is a number. In principle, the quadratic Hamiltonian can be exactly solved. With this method, the Kitaev model is exactly solved and produces the phase diagram, as shown in Fig. 16.11.

On this phase diagram there are two different phases. The first one is a gapped phase located at the corners of the parameter triangle. There are two types of excitations in this phase: fermions and vertices. When $J_x = J_y = 0$ and $J_z > 0$, the ground state of the system is just a set of dimers. Each dimer can be $|\uparrow\uparrow\rangle$ or $|\downarrow\downarrow\rangle$ independently and there are many degeneracies as shown in Fig. 16.12.

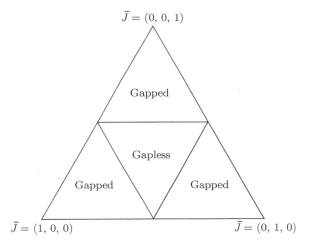

$$\bar{J} = (0, 0, 1)$$

Gapped

Gapless

Gapped   Gapped

$$\bar{J} = (1, 0, 0)$$ $$\bar{J} = (0, 1, 0)$$

Figure 16.11.   The phase of the Kitaev model.

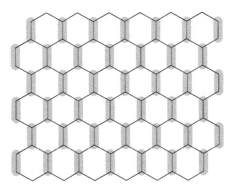

Figure 16.12.   The vertical dimers form the edge of the square lattice and the effective model of the Kitaev model can be reduced to the toric code model.

When $J_x, J_y \ll J_z$, the Hamiltonian can address the perturbation theory to obtain the 4th effective Hamiltonian as [43]

$$H_{\text{eff}} = -\frac{J_x^2 J_y^2}{16 J_z^3} \sum_p Q_p, \qquad (16.18)$$

where p is found over all the plaquettes of the dimer lattice and Q_p is the plaquette operator on the effective spin $|\uparrow\uparrow\rangle$ and $|\downarrow\downarrow\rangle$

$$Q_p = \sigma_{p_1}^y \sigma_{p_2}^x \sigma_{p_3}^y \sigma_{p_4}^x. \qquad (16.19)$$

This can effectively be transformed into the toric code model [37] by some modification of the effective spins. It is well known that the toric code model has two types of excitations, which are abelian anyons because these two types of particles cannot be commuted.

The other gapless phase is a spin liquid phase. A gap can be opened by a magnetic field, which breaks the time-reversal symmetry. Most interestingly, the perturbation will generate non-Abelian anyon excitations. These anyons are immune from the errors that can be used as quantum qubits and the braiding operation can be viewed as a unitary operator. Thus, quantum computation can be implemented in this system with high fault tolerance.

16.2.   Simulate spin systems in an optical lattice

16.2.1.   *Classical spin model*

The classical spin model gives some characteristics of the quantum spin model without quantum fluctuation. The difference between these two models is a very important phenomenon of the quantum system. For the phases that are insensitive to quantum fluctuation, the classical and quantum cases are the same. Thus, the classical spin model can also give us some insight into the quantum models.

The simplest way to create a frustrated system is with an antiferromagnetic interaction on a triangle lattice [38]. The antiferromagnetic interaction prevents the spins from aligning in the same direction. Even a classical spin system on the triangle lattice has rich phases. The classical model with a Heisenberg type interaction on the triangle lattice can be realized in an optical lattice.

For the weak interactions of ultracold bosonic atoms in an optical lattice, a superfluid state can be made very easily. Due to the coherence of the superfluid, the atoms at each site i of the lattice have a well-defined local phase θ_i. Every phase can be identified with a classical vector and the spin can be defined as $\bar{S}_i = (\cos(\theta_i), \sin(\theta_i))$. Thus, the local atoms can be viewed as a classical spin. The Hamiltonian of this system with a Heisenberg interaction can be expressed as

$$H = -\sum_{\langle i,j \rangle} J_{ij} \mathbf{S}_i \cdot \mathbf{S}_j = -\sum_{\langle i,j \rangle} J_{ij} \cos(\theta_i - \theta_j) = E(\theta_i), \qquad (16.20)$$

where $\langle i,j \rangle$ are the nearest neighbors. The ground state of this system minimizes $E(\theta_i)$ with parameters θ_i. The parameter J_{ij}s are defined in

the experiment as shown in Fig. 16.13, where there are just two different parameters (J and J') and they can be independently tuned from the
ferro- to antiferromagnetic regions. The experiment's results show that the
system will be in different spin configurations with different parameters, as
illustrated in Fig. 16.14. This is a six-site spin configuration, which is a part
of the larger lattice. With these spin configurations, the phase diagram can
be experimentally found (see Fig. 16.15).

The classical model gives us some basic concepts of the spin model, but
the quantum fluctuation is the key characteristic of the quantum system. It
is impossible to understand the spin liquid phase on this classic frustrated
system. We also need to realize a real quantum spin system in the optical
lattice to understand the spin liquid phase. However, it is currently very difficult to realize the spin models in the optical lattice due to the requirement
for very low temperatures. The following are several theoretical proposals
to realize the quantum spin models.

16.2.2. Quantum spin model

We have previously mentioned that there are several potential models whose
ground state is a spin liquid state. If we can realize these models in an optical

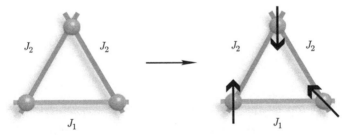

Figure 16.13.   The classical spin models on the triangle lattice where parameters J and
J' can be tuned.

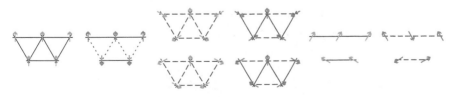

Figure 16.14.   Spin configuration on the triangle lattice.

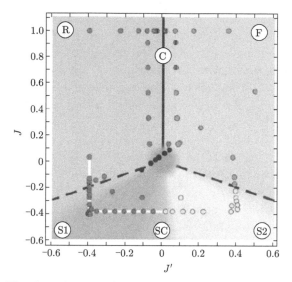

Figure 16.15.   The phase diagram of the classical spin model on a triangle lattice. The
phases are defined as follows: F: Ferromagnetic, C: 1D chains, R: Rhombic, S1: Spiral 1,
SC: staggered 1D chains and S2: Spiral. From Ref. [38].

lattice, then we can investigate the property of the spin liquid. In addition,
the other phases of the system can be investigated at the same time.

To simulate the spin-1/2, two different types of atoms [39] distinguishable by their internal states, are required to denote the effective spin
$\sigma = \uparrow, \downarrow$. The atoms with different spins $\sigma = \uparrow, \downarrow$ can be confined by an independent optical lattice with a different polarization or frequency. A peri-
odic potential is denoted by $V_{\mu\sigma} \sin^2(\vec{k}_\mu \cdot \vec{r})$ for the direction μ, where $\vec{k}_\mu$
is the wave vector of light. In each optical lattice, the system is similar to
the situation in the BH model. For a sufficiently strong periodic potential
and low temperature, the atoms will be in the lowest Bloch band. We can
also consider the Hamiltonian on the Wannier function basis in the lowest
band. For our purpose here, we need to control the hopping parameters
with field intensity, in order to create the ultracold atom system near the
MI phase with one atom at each site.

Under these conditions, the Hamiltonian of this system can be
expressed as [39]

$$H = -\sum_{\langle ij \rangle \sigma} \left(t_{\mu\sigma} a_{i\sigma}^\dagger a_{j\sigma} + \text{H.C.} \right) + \frac{1}{2} \sum_{i,\sigma} U_\sigma n_{i\sigma}(n_{i\sigma} - 1) + U_{\uparrow\downarrow} \sum_i n_{i\uparrow} n_{i\downarrow}.$$

$$(16.21)$$

There is a new variable to describe the collision interaction between different spins on the same site. The strength of this interaction is determined by the s-wave scattering radius between the different spins, denoted by $a_{s\uparrow\downarrow}$ and other parameters. In a cubic lattice, this interaction is

$$U_{\uparrow\downarrow} \approx \left(\frac{8}{\pi}\right)^{1/2} (ka_{s\uparrow\downarrow}) \left(E_r \bar{V}_{1\uparrow\downarrow} \bar{V}_{2\uparrow\downarrow} \bar{V}_{3\uparrow\downarrow}\right)^{1/4}, \qquad (16.22)$$

and

$$\bar{V}_{\mu\uparrow\downarrow} = \frac{4\bar{V}_{\mu\uparrow}\bar{V}_{\mu\downarrow}}{\left(\bar{V}_{\mu\uparrow}^{1/2} + \bar{V}_{\mu\downarrow}^{1/2}\right)^2} \qquad (16.23)$$

is the spin average potential in each direction, where $\mu = 1, 2, 3$ denote different directions. The other parameters are defined in the BH model. Furthermore, the system can be tuned to the deep MI region, that is, the parameters satisfy the condition $t_{\mu\sigma} \ll U_{\uparrow\downarrow}, U_\sigma$ and $\langle n_{i\uparrow} \rangle + \langle n_{i\downarrow} \rangle \simeq 1$. Under this condition, the interaction terms are dominant and the hopping terms can be considered as a perturbation. Then, the effective Hamiltonian of this system can be determined by the perturbation theory as

$$H = -\sum_{\langle i,j \rangle} \left[\lambda_{\mu z}\sigma_i^z\sigma_j^z \pm \lambda_{\mu\perp}(\sigma_i^x\sigma_j^x + \sigma_i^y\sigma_j^y)\right], \qquad (16.24)$$

where the effective spin operator is defined as

$$\sigma_i^z = n_{i\uparrow} - n_{i\downarrow}, \quad \sigma_i^x = a_{i\uparrow}^\dagger a_{i\downarrow} + a_{i\downarrow}^\dagger a_{i\uparrow}, \quad \sigma_i^y = -i(a_{i\uparrow}^\dagger a_{i\downarrow} - a_{i\downarrow}^\dagger a_{i\uparrow}) \qquad (16.25)$$

and the system parameters are defined as

$$\lambda_{\mu z} = \frac{t_{\mu\uparrow}^2 + t_{\mu\downarrow}^2}{2U_{\uparrow\downarrow}} - \frac{t_{\mu\uparrow}^2}{U_\uparrow} - \frac{t_{\mu\downarrow}^2}{U_\downarrow}, \quad \lambda_{\mu\perp} = \frac{t_{\mu\uparrow}t_{\mu\downarrow}}{U_{\uparrow\downarrow}}. \qquad (16.26)$$

So far, the interaction is the well-known anisotropic Heisenberg (XXZ) spin model. The energy scale of this model is t_σ^2/U_σ, which is very small due to the condition $t_{\mu\sigma} \ll U_{\uparrow\downarrow}, U_\sigma$. To observe the ground state of this spin model, we should make the thermal fluctuation of this system much lower than the energy scale of this model. However, this condition is still beyond current cooling technologies. In the previous calculation, we do not constrain the geometry of the lattice. With different parameters and different lattices, this method can be used to simulate different spin models.

Following the scheme proposed in Ref. [39], we need to construct a honeycomb lattice to simulate the Kitaev model. The effective $2D$ lattice in the x-y plane can be realized by raising the potential barriers along the vertical direction z in the 3D optical lattice. To create the hexagonal lattice, three trapping potentials are applied in the x-y plane and the forms of the potentials are

$$V_j(x,y) = V_0 \sin^2\left[k_\parallel(x\cos\theta_j + y\sin\theta_j) + \varphi_0\right], \tag{16.27}$$

where $\theta_1 = \pi/6$, $\theta_2 = \pi/2$ and $\theta_3 = -\pi/6$ corresponding to the lines in the figure. Each potential is constructed by two blue-detuned antipropagating laser beams above the x-y plane with an angle $\varphi_\parallel = 2\arcsin(1/\sqrt{3})$, such that the wave vector $k_\parallel$ projected onto the x-y plane has the value $k_\parallel = k/\sqrt{3}$. Then, by setting the initial phase $\varphi_0 = \pi/2$ and the maxima of these potentials will be located at the same position. Because of the blue detuning, the atoms will be confined at the minima of the lattice, which generates a honeycomb lattice as shown in Fig. 16.16. The lattice simultaneously traps atoms with different spin states.

So far, the spin character of the atoms has not been considered in the honeycomb lattice. Some spin-dependent potential is required to realize the Hamiltonian-like equation. The spin-dependent trapping potentials are generated in the x-y plane by applying blue-detuned standing-wave laser beams along the tunneling directions. The potentials are

$$V_{\nu\sigma}(x,y) = V_{\nu\sigma}\sin^2\left[k(x\cos\theta'_\nu + y\sin\theta'_\nu)\right], \tag{16.28}$$

Figure 16.16.   The minima of the potential form the honeycomb lattice.

where $\theta'_x = -\pi/3$, $\theta'_y = \pi$ and $\theta'_z = \pi/3$ of the special Kitaev model. All these interactions can be controlled independently along the different directions. The amplification $V_{\nu\sigma}$ should depend on the spin and directions (where the direction is denoted D_ν) as

$$V_{\nu\sigma} = V_{\nu+}|+\rangle_\nu\langle+| + V_{\nu-}|-\rangle_\nu\langle-|, \tag{16.29}$$

where $|+\rangle_\nu$ and $|-\rangle_\nu$ are eigenvectors of σ^ν with eigenvalues $+1$ and -1, respectively.

To realize the spin dependence of the potential for different directions, we should use the characteristics of the atoms to realize these potentials. Consider the case where the atomic energy levels are shown in Fig. 16.17. Here, $\sigma = \uparrow, \downarrow$ denote two hyperfine states with different energies. They can couple to the common level $|e\rangle$ through two blue-detuned laser beams L_1 and L_2, respectively. Δ is the detuning parameter. By the definition of the Pauli operator, the atoms are quantized along the z direction, which is perpendicular to the x-y plane. The laser beams L_1 and L_2 are phase-locked and polarized along the z direction. Along the D_z direction, it is sufficient to apply L_1 laser beam to induce the potential $V_{z\sigma}$. However, both lasers L_1 and L_2 with different relative phases should be applied to induce $V_{x\sigma}$ or $V_{y\sigma}$ for the direction D_x or D_y. The spin-dependent potential changes the intensity of the initial light field, but the minima of the light field do not change. The entire system can be well described by the Kitaev model with tunable parameters.

The Heisenberg antiferromagnetic model on the trimerized Kagome lattice can also be used to investigate the spin liquid state [40, 41]. The naive way to construct the Kagome lattice is based on the triangle lattice. The triangle lattice can be constructed by three standing waves: $\cos^2(\mathbf{k}_{1,2}\cdot\mathbf{r})$, where $\mathbf{k}_{1,2} = k(1/2, \sqrt{3}/2)$ and $\cos^2(\mathbf{k}_3 \cdot \mathbf{r} + \phi)$, where $\mathbf{k}_3 = k(0,1)$. When varying the initial phase ϕ, the third standing wave will shift along the y direction. Thus, the Kagome lattice will be constructed. However, there are some problems that prevent this naive construction from working [39]. Firstly,

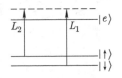

Figure 16.17. The level structure and the laser configuration used to induce spin dependent tunneling.

three standing waves cannot be mutually orthogonal because there are only two different polarizations on a plane. Thus, there will be undesired interference between the standing waves. Secondly, the more serious problem for this construction is that the Rayleigh limit keeps the minima of the laser intensity (minima for a red-detuned laser and maxima for a blue-detuned laser) from being resolved. Besides, the parameters along the same direction cannot be detuned with two periods, which is the characteristic of the trimerized Kagome lattice.

To overcome these problems, the lasers should be set up in a 3D configuration with a superlattice technique, as in Fig. 16.18. The arrows denote the wave vectors of the standing wave laser beams. The laser beams in the same plane are phase-locked and there are three vertical planes intersecting at 120°. The lasers in the same plane must have the same polarization and interference between the lasers in different planes must be avoided by introducing small frequency mismatches. With the experimental setup in the figure, there are three lasers in each plane. The intensity in the x-y plane will be

$$I(\mathbf{r}) = I_0 \sum_{i=1}^{3} \left[\cos\left(\mathbf{k}_i \cdot \mathbf{r} + \frac{3\sigma_i \phi}{2}\right) + 2\cos\left(\frac{\mathbf{k}_i \cdot \mathbf{r}}{3} + \frac{\sigma_i \phi}{2}\right) \right.$$
$$\left. + 4\cos\left(\frac{\mathbf{k}_i \cdot \mathbf{r}}{9} + \frac{\sigma_i \phi}{6}\right) \right]^2, \tag{16.30}$$

where $\mathbf{r} = (x, y)$, $\sigma_2 = -1$ and $\sigma_1 = \sigma_2 = 1$.

With this arrangement, the maxima of the optical lattice potential are well-resolved and correspond to those of the Kagome lattice. The resolution of the maxima can be quantitatively described by the parameter ξ; the ratio

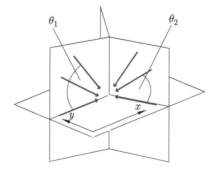

Figure 16.18. The laser configuration to induce a better Kagome lattice.

between the distance of the maxima and the half width at half maximum. In order to have a good resolution, the parameter ξ must be significantly bigger than 2. In the naive configuration, ξ is just about 4 to the ideal Kagome lattice. In this new setting, ξ is increased to about 14 and the lattice has a good resolution. In addition, the optical potentials can be smoothly transformed from an ideal Kagome lattice to the strongly trimerized lattice.

After the construct of the trimerized Kagome lattice, we put two fermion species in this lattice, using a method similar to that of the former case

$$H_{FF} = -\sum_{\langle ij \rangle} t_{ij} \left(f_i^\dagger f_j + \tilde{f}_i^\dagger \tilde{f}_j + \text{H.C.} \right) + \sum_i U n_i \tilde{n}_i, \qquad (16.31)$$

where f_j and $\tilde{f}_j$ are the fermion annihilation operators for the two species, respectively. The tunneling matrix elements are $t_{ij} = t$ for intratrimer and $t_{ij} = t'$ for intertrimer nearest-neighbor tunneling. H_{FF} is then the spin-1/2 Hubbard model. In the strong coupling limit $(t, t') \ll V$, this model can be transformed into the t-J model which reduces to the spin-1/2 Heisenberg model at half filling, that is

$$H_{FF} \to H_{\text{Heisenberg}} = J \sum_{\langle ij \rangle} \vec{\sigma}_i \cdot \vec{\sigma}_j + J' \sum_{\langle ij \rangle} \vec{\sigma}_i \cdot \vec{\sigma}_j, \qquad (16.32)$$

where the spin operators are defined as

$$n - \tilde{n} = 2\sigma^x, \quad f^\dagger \tilde{f} = \sigma^x + i\sigma^y, \qquad (16.33)$$

and the parameters are defined as $J = 4t^2/U$ and $J' = 4t'^2/U$. The energy scale of this model is determined by the scale of J and J'. Considering the experimental difficulties of cooling fermions, the temperature condition cannot be realized with the present technology.

The former proposed spin model are very difficult to realize in atomic systems due to the very low necessary temperature. The primary obstacle is that the super-exchange interaction, t^2/U, is quite weak. The only experimentally realized quantum spin model in optical lattice is the quantum Ising chains. By some subtle insights, this model can be mapped from the hopping interaction with field gradients, thus avoiding the weak super-exchange interaction [42].

The 1D BH model with a gradient magnetic field is

$$H = -t \sum_j \left(b_j^\dagger b_{j+1} + b_j b_{j+1}^\dagger \right) + \frac{U}{2} \sum_j \hat{n}_j (\hat{n}_j - 1) - \sum_j E_j \hat{n}_j. \quad (16.34)$$

For the parameter $U \gg t$, the system will be in a MI phase with N atoms in each site. With the addition condition $U = E$, the atoms on the site j can tunnel to the neighbor site $j-1$ or $j+1$ without energy cost, when the number of atoms on the neighbor site are the same.

The tunneled atom and a local atom can generate a dipole excitation. To begin with, we define a dipole creation operator on the site j as

$$d_j^\dagger = \frac{b_j b_{j+1}^\dagger}{\sqrt{N(N+1)}}. \quad (16.35)$$

Then the dipole operator satisfies the conditions

$$d_{j+1}^\dagger d_{j+1} d_j^\dagger d_j = 0, \quad (16.36)$$

With the aid of these dipole operators, the tilt BH model can be transformed into the form

$$H = -\sqrt{N(N+1)} t \sum_j (d_j^\dagger + d_j) + (U - E) \sum_j d_j^\dagger d_j. \quad (16.37)$$

Furthermore, we define a link with (without) a dipole excitation as the spin up (down) state along the z direction. The dipole operator d_j plays the same role as the flip operator on the spin j, as illustrated in Fig. 16.19. More definitely, we set

$$S_z^j = \frac{1}{2} - d_j^\dagger d_j,$$

$$S_x^j = \frac{1}{2}(d_j^\dagger + d_j),$$

$$S_y^j = \frac{i}{2}(d_j^\dagger - d_j). \quad (16.38)$$

The main insight of this system is that the constraint, $d_{j+1}^\dagger d_{j+1} d_j^\dagger d_j = 0$, can be implemented by adding a new term $J d_{j+1}^\dagger d_{j+1} d_j^\dagger d_j$ into the Hamiltonian. This process is just like the minimum-maximum problem, but with conditions. The parameter J plays the role as a Lagrange multiplier with

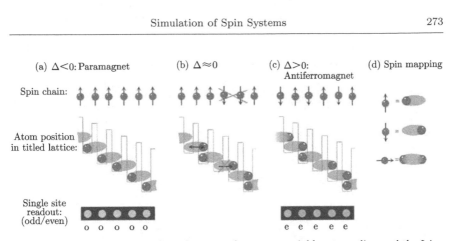

Figure 16.19.   The correspondence between the nearest neighbor tunneling and the Ising spin model. From Ref. [42].

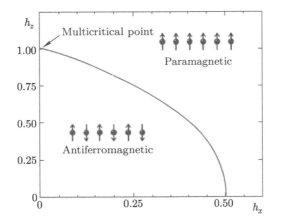

Figure 16.20.   The phase diagram of the Ising model with a magnetic field. From Ref. [42].

the order of U. Then the whole system with constraints can be described by the following spin Hamiltonian with $\Delta = E - U$

$$H = J \sum_j S_z^j S_z^{j+1} - 2\sqrt{N(N+1)}t \sum_j S_x^j - (J - \Delta) \sum_j S_z^j,$$

$$= J \sum_j \left(S_z^j S_z^{j+1} - h_x S_x^j - h_z S_z^j \right), \qquad (16.39)$$

where $h_x = 2^{3/2}t/J = 2^{3/2}\tilde{t}$ and $h_z = (1 - \Delta/J) = 1 - \tilde{\Delta}$ when $N = 1$. With the condition $\Delta \sim 0$, the parameters h_x and h_z are located near $(0,1)$. The phase diagram of such a spin model is given in Fig. 16.20.

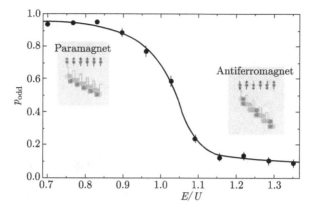

Figure 16.21.   Phase diagram of a quantum Ising model with a magnetic field. From Ref. [42].

In this model, the energy scale U is much bigger than the super-exchange t^2/U. This quantum spin model has been realized in an experiment [42]. In this experiment, the spin state was measured with site-by-site number resolution technology. This technology can give information on the number of atoms on a site: whether it is an odd or even number. If the number is odd, the phase should be paramagnetic. Conversely, if the number is even, the system should be anti-paramagnetic. In Fig. 16.21, the experimental results are shown in comparison to theoretical predictions, where a quantitative agreement is observed.

References

[1] K. Binder. Ising Model, in *Encyclopedia of Mathematics*. Ed. M. Hazewinkel. Springer, Berlin (2001).
[2] L. Onsager. *Phys. Rev.* **65**, 117 (1944).
[3] R. J. Baxter. *Exactly Solved Models in Statistical Mechanics*. Academic Press, London (1982).
[4] M. Lewenstein, A. Sanpera, V. Ahufinger, B. Damski, A. Sen(De) and U. Sen. *Adv. Phys.* **56**, 243 (2007).
[5] G. Misguich and C. Lhuillier, in *Frustrated Spin Systems*. Ed. H. T. Diep. World Scientific, Singapore (2004).
[6] C. Lhuillie. arXiv: cond-mat/0502464.
[7] C. K. Majumdar and D. K. Ghosh. *J. Math. Phys.* **10**, 1388 (1969).

[8] I. Affleck, T. Kennedy, E. H. Lieb and H. Tasaki. *Commun. Math. Phys.* **115**, 477 (1988).

[9] M. Fannes, B. Nachtergaele and R. F. Werner. *Commun. Math. Phys.* **144**, 3 (1992).

[10] S. Ostlund and S. Rommer. *Phys. Rev. Lett.* **75**, 19 (1995).

[11] A. Auerbach. *Interacting Electrons and Quantum Magnetism.* Springer, New York (1994).

[12] T. Senthil, A. Vishwanath, L. Balents, S. Sachdev and M. P. A. Fisher. *Science* **303**, 1490 (2004).

[13] A. Kitaev. *Ann. Phys.* **321**, 2 (2006).

[14] P. W. Anderson. *Science* **237**, 1196 (1987).

[15] C. Waldtmann, H.-U. Everts, B. Bernu, C. Lhuillier, P. Sindzingre, P. Lecheminant and L. Pierre. *Eur. Phys. J. B* **2**, 501 (1998).

[16] C. Zeng and V. Elser. *Phys. Rev. B* **42**, 8436 (1990).

[17] R. Singh and D. Huse. *Phys. Rev. Lett.* **68**, 1766 (1992).

[18] P. Leung and V. Elser. *Phys. Rev. B* **47**, 5459 (1993).

[19] C. Zeng and V. Elser. *Phys. Rev. B* **51**, 8318 (1995).

[20] P. Sindzingre, P. Lecheminant and C. Lhuillier. *Phys. Rev. B* **50**, 3108 (1994).

[21] T. Nakamura and S. Miyashita. *Phys. Rev. B* **52**, 9174 (1995).

[22] P. Lecheminant, B. Bernu, C. Lhuillier, L. Pierre and P. Sindzingre. *Phys. Rev. B* **56**, 2521 (1997).

[23] F. Mila. *Phys. Rev. Lett.* **81**, 2356 (1998).

[24] M. Mambrini and F. Mila. *Eur. Phys. J. B* **17**, 651 (2001).

[25] R. Budnik and A. Auerbach. *Phys. Rev. Lett.* **93**, 187205 (2004).

[26] S. Yan, D. A. Huse and S. R. White. *Science* **332**, 1173 (2011).

[27] T.-H. Han, J. S. Helton, S. Chu, D. G. Nocera, J. A. Rodriguez-Rivera, C. Broholm and Y. S. Lee. *Nature* **492**, 406 (2012).

[28] N. Read and S. Sachdev. *Phys. Rev. Lett.* **62**, 1694 (1989).

[29] N. Read and S. Sachdev. *Phys. Rev. Lett.* **66**, 1773 (1991).

[30] G. Murthy and S. Sachdev. *Nucl. Phys. B* **344**, 557 (1990).

[31] V. N. Kotov, O. P. Sushkov, J. Oitmaa and Z. Weihong. *Phys. Rev. B* **60**, 14613 (1999).

[32] M. P. Gelfand, R. R. P. Singh and D. A. Huse. *Phys. Rev. B* **40**, 10801 (1989).

[33] M. P. Gelfand. *Phys. Rev. B* **42**, 8206 (1990).

[34] R. R. P. Singh, Z. Weihong, C. J. Hammer and J. Oitmaa. *Phys. Rev. B* **60**, 7278 (1999).

[35] L. Wang, Z.-C. Gu, F. Verstraete and X.-G. Wen. arXiv:1112.3331.

[36] H. C. Jiang, H. Yao and L. Balents. *Phys. Rev. B* **86**, 024424 (2012).

[37] A. Y. Kitaev. *Ann. Phys.* **303**, 2 (2003).

[38] J. Struck, C. Olschlager, R. Le Targat, P. Soltan-Panahi, A. Eckardt, M. Lewenstein, P. Windpassinger and K. Sengstock. *Science* **333**, 996 (2011).

[39] L.-M. Duan, E. Demler and M. D. Lukin. *Phys. Rev. Lett.* **91**, 090402 (2003).

[40] L. Santos, M. A. Baranov, J. I. Cirac, H.-U. Everts, H. Fehrmann and M. Lewenstein. *Phys. Rev. Lett.* **93**, 030601(2004).

[41] B. Damski, H. Fehrmann, H.-U. Everts, M. Baranov, L. Santos and M. Lewenstein. *Phys. Rev. A* **72**, 053612 (2005).

[42] J. Simon, W. S. Bakr, R. Ma, M. E. Tai, P. M. Preiss and M. Greiner. *Nature* **472**, 307 (2011).

[43] A. Kitaex and C. Laumann. Topological phases and quantum computation, in *Exact Methods in Low-dimensional Physics and Quantum Computing*. Les Houches Summer School (2008).

CPSIA information can be obtained
at www.ICGtesting.com
Printed in the USA
LVHW081527040119
R14411500001B/R144115PG600626LVX4B/1/P

9 789813 270756